City Sustainability and Regeneration

WITPRESS

WIT Press publishes leading books in Science and Technology.
Visit our website for the current list of titles.
www.witpress.com

WITeLibrary

Home of the Transactions of the Wessex Institute.
The WIT electronic-library provides the international scientific community with immediate
and permanent access to individual papers presented at WIT conferences.
http://library.witpress.com

City Sustainability and Regeneration

Editors

S. Mambretti
Polytechnic of Milan, Italy

J. L. Miralles i Garcia
Politechnic University of Valencia, Spain

WITPRESS Southampton, Boston

Editors:

S. Mambretti
Polytechnic of Milan, Italy

J. L. Miralles i Garcia
Politechnic University of Valencia, Spain

Published by

WIT Press
Ashurst Lodge, Ashurst, Southampton, SO40 7AA, UK
Tel: 44 (0) 238 029 3223; Fax: 44 (0) 238 029 2853
E-Mail: witpress@witpress.com
http://www.witpress.com

For USA, Canada and Mexico

Computational Mechanics International Inc
25 Bridge Street, Billerica, MA 01821, USA
Tel: 978 667 5841; Fax: 978 667 7582
E-Mail: infousa@witpress.com
http://www.witpress.com

British Library Cataloguing-in-Publication Data

A Catalogue record for this book is available
from the British Library

Library of Congress Catalog Card Number: 2020932215

ISBN: 978-1-78466-415-2
eISBN: 978-1-78466-416-9

The texts of the papers in this volume were set individually by the authors or under their supervision. Only minor corrections to the text may have been carried out by the publisher.

The material contained herein is reprinted from the following Journals, which are/were all published by WIT Press: the *International Journal of Transport Development and Integration*, Volume 4(**1**); the *International Journal of Design and Nature and Ecodynamics*, Volumes 14(**2**, **3**); the *International Journal of Energy Production and Management*, Volume 4(**4**) and Volume 5(**1**); the *International Journal of Environmental Impacts*, Volume 3(**2**).

Preface

A set of new studies are included in this volume which provides solutions that lead towards sustainability. Contributions originate from a diverse range of researchers, resulting in a variety of topics and experiences.

Urban areas face a number of challenges related to reducing pollution, improving main transportation and infrastructure systems and these challenges can contribute to the development of social and economic imbalances and require the development of new solutions. The challenge is to manage human activities, pursuing welfare and prosperity in the urban environment, whilst considering the relationships between the parts and their connections with the living world. The dynamics of its networks (flows of energy matter, people, goods, information and other resources) are fundamental for an understanding of the evolving nature of today's cities.

Large cities represent a productive ground for architects, engineers, city planners, social and political scientists able to conceive new ideas and time them according to technological advances and human requirements. The multidisciplinary components of urban planning, the challenges presented by the increasing size of cities, the amount of resources required and the complexity of modern society are all addressed.

The Editors

Contents

WOULD A SUSTAINABLE CITY BE SELF-SUFFICIENT IN FOOD PRODUCTION?

GASTON E. SMALL[1,*], ROBERT MCDOUGALL[2] & GENEVIÈVE SUZANNE METSON[3]
[1] Biology Department, University of Saint Thomas, Saint Paul, Minnesota, United States.
[2] School of Environmental and Rural Sciences, University of New England, Armidale, NSW, Australia.
[3] Department of Physics, Chemistry, and Biology (IFM), Linköping University, Linköping, Sweden.

ABSTRACT
Urban agriculture has increased in many cities and has the potential to provide an array of benefits including increased local food production, nutrient recycling, urban green space, and biodiversity. While certain environmental benefits of urban agriculture are evident, it is not clear what the optimal extent of urban agriculture would be in designing a sustainable city. Closing the loop by recycling waste products into new resources is fundamental to sustainability, but the extent to which this should occur at local, regional, or global scales is an open question. We analyze how potential benefits and costs associated with urban agriculture scale with the extent of implementation, and compare potential tradeoffs in different metrics of sustainability. We assess how the appropriate metrics to optimize in a given city are context-dependent. For example, maximizing production in a small land footprint could be important in densely developed urban environments, whereas filling vacant land with food-producing gardens may be a more appropriate goal in certain post-industrial cities. We assess the potential role that urban agriculture plays in making urban food systems more resilient to climate change and other disruptions. Finally, we consider a case study comparing the resources required and pollution generated to produce the lettuce supply of on U.S. metropolitan area through outdoor urban agriculture and indoor urban agriculture, compared to conventional production and cross-continental transportation. This analysis illustrates the importance of considering multiple metrics in assessing sustainability of urban agriculture.
Keywords: cost-benefit analysis, sustainability, resilience, trade-offs, urban agriculture.

1 INTRODUCTION

'A good way to see the embedding problem is to imagine the consequences of cutting off all flows in and out, as military sieges of European castles and cities attempted to do in the Middle Ages. From this point of view and in the short term of days to months, some farms and ranches would be reasonably sustainable, but the residents of a large city or an apartment building would rapidly succumb to thirst, starvation, or disease. Viewed from this perspective, even though Portland may be the greenest and by some definitions 'the most sustainable city in America', it is definitely not self-sustaining.' [1]

The quote above highlights the degree to which cities are open ecosystems, depending on imports of food and other materials, and exportation of waste. A city cut off from its food supply, whether by an invading army or a natural disaster, quickly descends into chaos. This passage also raises questions about the relationship between self-sufficiency and sustainability. Would a sustainable city be self-sufficient in food production? Should urban planners strive towards maximizing urban food production in the name of sustainability, or might economic, social, and environmental costs exceed benefits at some point? While the recent growth of urban agriculture (UA) has provided social and environmental benefits, resources required (e.g. space, labor, water, energy) and pollution generated (e.g. nutrient-laden runoff) could eventually present barriers to continued expansion. These potential tradeoffs add complexity to understanding the role of urban agriculture in sustainability.

---

* ORCID: *http://orcid.org/0000-0002-9018-7555*

© 2020 WIT Press, www.witpress.com
DOI: 10.2495/DNE-V14-N3-178-194

Cities generally require a total land area that is 200-300 times larger than the geographic footprint of the city itself to provide food and other resources, and to assimilate wastes [2]. Historically in the United States, much food production for cities occurred in the surrounding hinterlands, but the industrialization of agriculture in the 20th Century and growth of transportation infrastructure has led to long supply chains. One study estimated the average direct transport distance of food in the United States to be 1,640 km, and this distance stretches to over 6,000 km if the entire supply chain is taken into account [3]. The long-distance movement of food has added to the environmental footprint of food production through raising transportation-related CO_2 emissions [4]. However, the majority of greenhouse gas emissions are associated with crop production rather than transportation [3], and a nationwide or global food production network allows crops to be grown in areas where resource supply (climate, water, labor) is favorable. Local production may not be inherently more sustainable, despite common assumptions [5].

The expansion of UA in recent years provides a range of potential social, economic, health, and environmental benefits to urbanites [6]. UA has potential environmental benefits including reduction of transportation-related energy consumption [7], creating habitat for pollinators [8], reducing urban heat island effect [9], and providing a beneficial reuse for wastewater and organic matter [10, 11]. Potential social benefits include strengthening connections between farmers and consumers [12], connecting urban residents with nature [13, 14], and improving livability of cities [15–17]. Potential public health benefits include lowering disease risk due to adoption of more plant-based diets [18, 19], as well as benefits to general well-being [20, 21]. Potential economic benefits of UA include stimulating green-sector employment [22] and improving food access to low-income residents [23].

However, many of these benefits are also coupled with challenges [6]. For example, while UA may create green jobs, it often relies on underpaid labor, and commercial UA jobs that earn a living wage are rare relative to the number of students who are being trained for potential careers in this field [23]. While UA may increase urban property values in some localities [24], other studies have found limited economic impacts [25]. While UA can reduce greenhouse gas emissions related to transportation and storage [26], it may lead to increased emissions from climate-controlled local food production [27]. While UA can result in increased urban nutrient recycling through the use of compost, nutrient recycling efficiency may often be very low [28, 29] resulting in nutrient loss to groundwater and surface water [30, 31]. While UA systems may provide a more valuable habitat for urban biota than lawns or other common urban land uses [32], urban wildlife populations may be too depauperate to take advantage of this improved habitat [33]. Land used for UA may have other beneficial uses, presenting opportunity costs for local food production. Furthermore, different types of UA (e.g. backyard gardens, rooftop gardens, vertical farming) require different resources and provide different benefits [6]. While the cumulative potential value of ecosystem services provided by UA has been estimated as high as 160 billion USD [34], the exact nature of these benefits (and costs) is highly dependent on the context of a particular city [6, 35].

2 OBJECTIVES AND METHODOLOGY

In order to evaluate the optimal extent of UA in a given urban area, a holistic analysis of social and environmental benefits and costs is required. In this paper, we first evaluate the production potential of urban agriculture by compiling values reported from various case studies. We examine potential metrics of sustainability in UA and explore potential trade-offs among these metrics. We present a conceptual model as a framework for considering which metrics of sustainability should be optimized in various contexts. We analyze how potential benefits and costs associated with UA scale with the extent of implementation, and we evaluate the potential role that UA may play in making urban food systems more resilient by reviewing relevant case studies. Finally, we present an original case study examining the economic and environmental costs of producing the lettuce supply for Minneapolis-Saint Paul, Minnesota, from either outdoor or indoor UA, compared to importation from central California.

3 PRODUCTION POTENTIAL OF URBAN AGRICULTURE

For UA production to fulfill global urban vegetable demand would require one-third of all urban spaces [36], although this analysis does not account for the likely extensive variation among cities. Many cities in mild climate zones have the potential for UA to fulfill a substantial fraction of urban fruit and vegetable demand under intensive outdoor production scenarios (Fig. 1). However, most urban crops are generally low in calories so that even these high-production scenarios are not contributing a significant amount of food calories. There is precedent for urban vegetable self-sufficiency: the Paris market gardens of the late 19th Century covered ⅙ of the city's land area and provided more than 100% of the city's demand for salad crops (but only 1.4% of caloric requirements, and 2.4% of protein requirements [37]). If peri-urban agriculture is included, the capacity for local self-sufficiency increases considerably. Twenty-one percent of U.S. metropolitan statistical areas (MSAs) are currently capable of local self-sufficiency for milk and eggs, 12% of MSAs are capable of local self-sufficiency for fruit, and 16% of MSAs are capable of local self-sufficiency for vegetable production [38].

If the sustainability of UA is to be fully assessed, the assumptions underlying these production scenarios should be critically examined. For example, several UA production scenarios involve extensive use of industrial and commercial rooftops [39–41]. In some cases, rooftops would need to be retrofitted to handle the additional weight [40], which could pose a significant expense, and the potential need for irrigation and potential nutrient pollution from

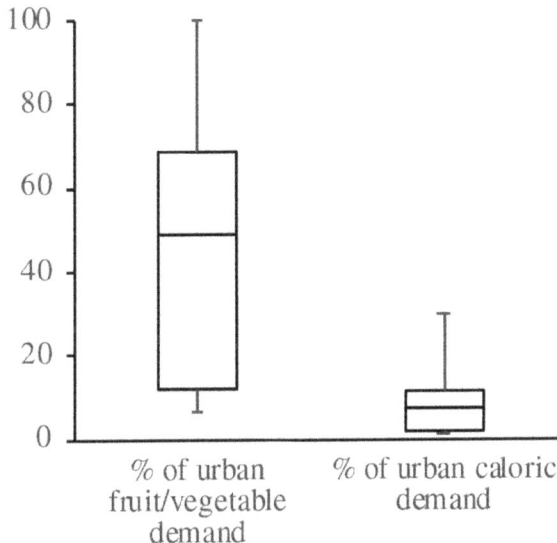

Figure 1: Percent of urban fruit and vegetable production demand (n = 8 cities), and percent of urban caloric demand (n = 12 cities), potentially supplied by urban agriculture under intensive outdoor production scenarios. Data are from [6, 39, 40, 90–100].

rooftop runoff [42] would need to be weighed against the economic, social, and environmental benefits of this expanded urban food production. Other intensive UA production scenarios assume extensive cultivation of available vacant land and some fraction of residential land [6, 39]. Under these scenarios, the intensive human labor requirement [43] and the potential for nutrient pollution from compost application [31] should be considered along with the direct benefits of food production plus indirect benefits such as providing pollinator habitat, fostering civic engagement, and promoting public health [6]. While proponents of indoor vertical farms offer a more expansive vision of urban food production, such as 30-story farms on a city block that could feed 50,000 people with vegetables, fruit, eggs, and meat [44], evidence supporting the economic and environmental feasibility of this vision are lacking.

4 TRADEOFFS AMONG METRICS OF SUSTAINABILITY

Many UA practitioners cite environmental benefits or sustainability as a key motivating factor [43]. UA has been characterized as resource-efficient based on use of vacant space, potential use of stormwater for irrigation, and reduction of food miles [45]. However, UA can be inefficient in terms of human labor [43] and nutrient recycling efficiency [29, 31]. Assessing the environmental sustainability of UA requires consideration of a wide variety of metrics, such as: biomass yield per unit area; calories or protein yield per unit area; yield per human energy input; yield per fossil fuel energy input; yield per unit input of nutrients or water; yield per pollution generated; output value relative to input value; net profit per unit area; net profit relative to capital inputs; the use of recycled nutrients; and the provisioning of pollinator habitat or other ecosystem services.

It is likely that there are tradeoffs among some of these metrics. For example, maximizing yield per unit area may require additional inputs of water, fertilizer, and labor, which could reduce the efficiency in terms of those metrics. Maximizing profitability may require production of fast-growing leafy greens or herbs, rather than in crops higher in calories or protein. Indoor production can achieve high yields and high efficiency of space, water, and nutrient use, but this type of UA typically requires high capital inputs and does not recycle nutrients from urban waste or provide habitat for pollinators and other urban biota. Determining which metric to optimize is context-dependent, based on resource availability in a given city, or for a given producer. In a densely populated urban area such as Brooklyn, optimizing the spatial footprint of UA (e.g. through rooftop farming) may be a top priority, whereas in post-industrial cities such as Cleveland or Detroit, where vacant lots are abundant, space may not be a constraint. In water-stressed cities such as Los Angeles, optimizing water use in UA may be a higher priority than in cities like Seattle that receive ample rainfall. Sustainability requires optimizing the use of the scarcest resource.

A conceptual framework for evaluating sustainability of UA is shown in Fig. 2. Urban agriculture creates the potential for local food production fueled by recycled organic waste and water. Like all forms of agriculture, UA requires inputs of energy, labor, space, water, and nutrients, and generates desired outputs (crop yield) as well as undesired outputs (pollution). The availability of these resources is represented by the cost to UA practitioners, or, for some resources in the case of outdoor UA production (sunlight, ambient rainfall, temperature range), is determined by climate zone. Inputs required for UA such as nutrients and water may have high amounts of embedded energy, or fossil fuel used in the production chain of these resources [46]. The environmental impact of nutrient pollution from urban gardens depends on the capacity of the local soils to retain (e.g. through adsorption of phosphorus) and remove (e.g. through microbial denitrification) these nutrients, and the sensitivity of downstream aquatic ecosystems that would receive these nutrients. There are both opportunities to

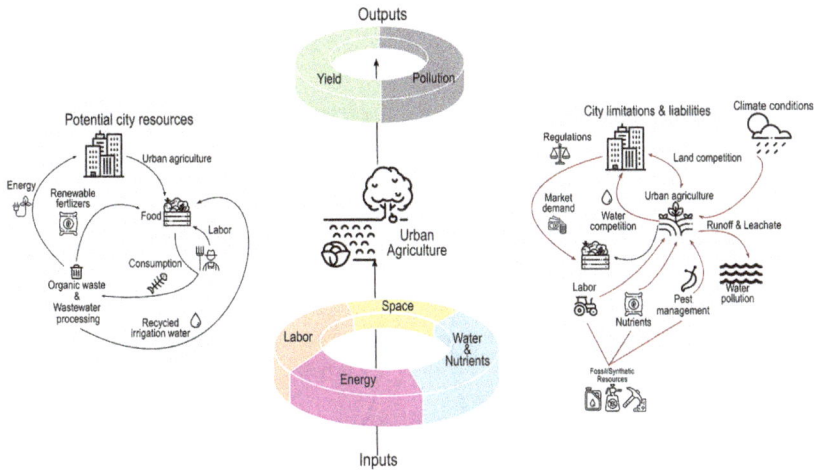

Figure 2: Conceptual diagram for evaluating sustainability of Urban Agriculture (UA), focused on material and economic flows. Center panel: UA requires inputs of labor, energy, space, water, and nutrients, and generates both desired outputs (crop yield) and undesired outputs (pollution). UA presents potential opportunities to sustainably acquire inputs and manage outputs (left circle), but also poses multiple constraints and potential sustainability trade-offs (right circle).

sustainably acquire inputs and manage outputs, but cities also face multiple constraints and potential sustainability goal trade-offs that can arise from UA. UA creates potential to help close the material cycling loop in cities through use of recycled nutrients and water to provide food, and be compatible with green energy harvesting from organic waste. It can also provide other circular economy benefits like employment. However, not all cities have the space, money, or growing conditions to support intensive UA without damaging social and ecological resources.

The economic feasibility of UA can also be assessed through this framework. While net economic benefit may be of small concern to many home gardeners [6, 43], larger-scale UA (i.e. urban farms) must be profitable to exist, unless supported by cities, universities, or other entities. Gross revenue from crop production depends on crop yield and crop value, which depends on the local market. Scaling up local food production could saturate market demand and depress prices. Production costs include costs of inputs (fertilizer, water, pesticide, energy for indoor growing), labor, and the cost of processing, transporting, and distributing crops. Additionally, the cost of land/space and any capital costs for equipment must be factored in.

Different forms of UA generally require different amounts of resources and generate different amounts of yield and pollution. Outdoor UA utilizing raised beds is one of the most common types of UA [47]. Yield can be high relative to conventional rural agriculture due to intercropping and planting at a higher density than mechanized agriculture allows for, but human labor inputs are also very high [37, 43]. Non-solar energy inputs are minimal, and required supplemental water inputs depend on local climate. However, there is potential for nutrient pollution due to runoff or leachate, depending on the type and quantity of compost or other soil amendments applied [31, 28]. Capital cost is generally very low, and this type of UA can be implemented on residential lawns or vacant lots, so space may not be a limitation except in high-density urban developments (Fig. 3).

**Urban
Agriculture
(outdoor raised bed)**

**Urban
Agriculture
(indoor vertical farm)**

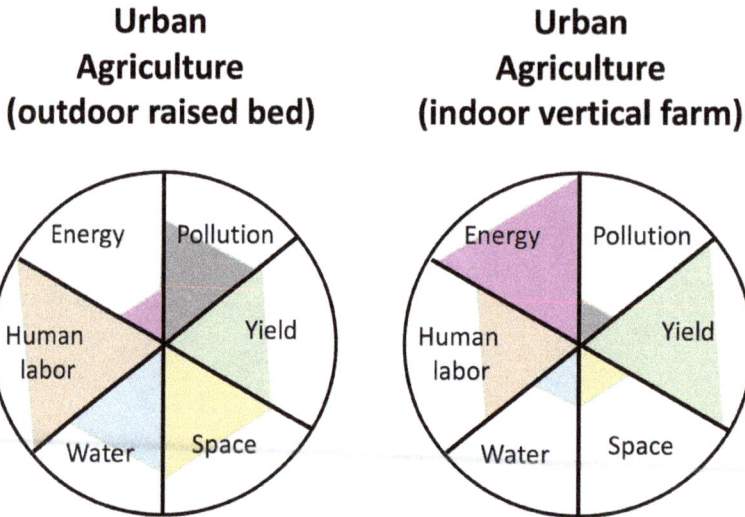

Figure 3: Relative magnitudes of inputs (Energy, Human Labor, Water, Space) and outputs (Yield, Pollution) from outdoor and indoor urban agriculture.

In contrast, indoor vertical farms typically use hydroponics systems and artificial light to grow crops, and can achieve extremely high yields per unit area by growing up to 8 crop cycles per year [48] and stacking multiple vertical layers of growing beds. One hectare of hydroponic greenhouse production has the potential to replace 10 hectares of rural land [41]. Commercial tomato production in greenhouses can achieve yields up to 15 times higher than outdoor growing [49], and indoor farms or rooftop greenhouses in New York City report yields of lettuce greens that are approximately 30 times higher compared to conventional outdoor production [23]. Water use in indoor growing systems is highly efficient, 75% lower than conventional farming [50]. Indoor production systems have the potential to recycle water transpired by crops [51], harvest rainwater [50], or utilize graywater [52]. Most indoor UA production systems have a high non-solar energy requirement for heat and lighting, although the use of waste heat from buildings can reduce this requirement [53]. However, the amount of CO_2 generated per kg of vegetable produced may be 2-5x greater than for produce grown outdoors [4, 27], exceeding CO_2 emissions from cross-country transport of crops [54]. Human labor required may vary depending on the degree of mechanization of the indoor growing operation [23]. While hydroponics systems may not be a source of nutrient pollution, they also do not usually recycle waste nutrients, instead relying on inputs of industrially derived nutrients to maintain high yields [55]. Indoor UA production also does not provide environmental benefits such as pollinator habitat, or social benefits of urban residents being in contact with gardens. Because of the high capital requirements [56] and complex profit model [57] of commercial indoor production, production tends to focus on salad greens and herbs that can be sold for premium prices, and therefore are not accessible to many low- and middle-income consumers [23].

5 SCALE AND SUSTAINABILITY

In addition to differences between indoor and outdoor UA systems, the scale of UA production is also likely to affect metrics of sustainability. For example, human labor efficiency may be extremely low in backyard gardening, where production and efficiency are generally not motivating factors of these gardeners [43], however some of this apparent inefficiency may be due to gardeners mixing recreational activities with labor [58]. For larger scale commercial urban farms, efficient use of labor may be necessary to maintain viability. UA is generally able to achieve higher yields per unit area compared to conventional agriculture due to inter-cropping and polyculture, which depends on human labor. At larger scales (where labor, rather than space, becomes a limiting factor), mechanization replaces human labor and necessitates a shift from polyculture to monoculture [59].

Nutrient use efficiency is likely to increase with increasing scale. For backyard gardeners, there is little economic incentive to be judicious in use of fertilizer or compost, and extremely high fertilizer use has been documented in some urban food gardens [60]. For larger area operations, the cost and labor required to apply fertilizer or other soil amendments creates incentives to actively manage soil fertility in response to soil test results. This relationship is illustrated by the relatively high nitrogen-use efficiency (20–70% [61]) and phosphorus-use efficiency (60–100% [62]) for conventional agriculture, compared to values below 5% that have been documented for outdoor, soil-based UA [29, 31, 28]. The use of other resources such as water is also likely to become more efficient in larger scale production systems.

However it is worth noting that, depending on the source of the material in question, higher levels of some inputs could in fact improve the sustainability of UA systems. This could be the case with amendments based on organic waste material, such as compost or mulch. Organic material, such as food scraps and garden waste, forms a major part of the domestic waste stream of many cities in developed nations [63], much of it going to landfill, resulting in the emission of greenhouse gases both as a result of its management and decomposition. It could thus be argued that increasing inputs of these materials into UA could be of benefit to the environment and economy of the city in which it occurs, even if they are in excess of what is needed, so long as they don't exceed the ecological carrying capacity of the system. The same is true of water, which, if captured as run-off from impervious surfaces, can reduce erosion and contamination of nearby water bodies [64].

6 URBAN AGRICULTURE AND RESILIENCE OF URBAN FOOD SYSTEMS
In times of crisis UA has provided a number of direct benefits to urban citizens in terms of access to food and green space, but also continues to play a role in collective memories and skills about how to grow food and create communities [65]. For example, Stockholm's allot-ment gardens participants have meaningful communities of practice where people learn from each other about ecosystem services and reflexively adapt to changing circumstances [66]. The expansion of UA has potential to make urban food systems more resilient, by reducing reliance on external imports during times of scarcity. For example, in the United States during World War II, half of all families planted Victory Gardens, producing 55 kg of fruit and vegetables annually for every civilian [67]. In Cuba, following the fall of the Soviet Union, the caloric intake of Cubans dropped nearly 30%, and the government instituted a goal of cultivating 10 m^2 of urban land for every resident. By 2000, Cuba surpassed pre-crisis levels of food production, and these urban farms provided up to 50% of caloric intake and reduced the need for imported fuel to distribute food and generate electricity for refrigeration [68]. Beyond political disasters, climate change, and especially extreme weather events, can threaten food supply chains [69]. Local production from UA can serve as a buffer to climate-related disruptions [70] or other external disruptions, such as sharp increase in fossil fuel costs. On the other hand, the effects of climate change on UA production could be positive (e.g. by increasing the length of the outdoor growing season in some cooler-climate cities) or negative (e.g. by increasing frequency of

severe droughts in some areas). If climate change leads to increasing scarcity of water in some cities, then the cost (both economic and environmental) of UA would increase correspondingly. While UA may help decrease urban heat island effects through evaporative cooling [71], water is the currency required for this ecosystem service, and climate change may drive this cost to increase in the future. For example, recent research in Melbourne, Australia, has found that urban gardeners have increased water use to adapt to increased temperatures (related to climate change and exacerbated by the urban heat island effect), which could be a maladaptation if there is water scarcity [72]. On the other hand, inefficiencies in water use that might occur when practicing UA can offer a 'buffer' when resources are scarce, providing resilience in terms of food, water, and even energy as in past civilizations like the Mayan and Byzantine empires [73].

It is also possible that increased reliance on UA could lead to potential vulnerabilities in food supply. Outdoor production could be susceptible to extreme weather events and pest outbreaks. Indoor production is susceptible to disruptions from power outages. There have been numerous documented instances of short-term power outages destroying aquaponics or hydroponics crops [74–76]. Backup generators provide a buffer to short-term power disruptions, but during extended disruptions (e.g. power outages associated with Hurricane Maria in Puerto Rico in 2017), there may be more pressing needs for fuel and electricity than keeping plants or fish alive.

7 CASE STUDY: LETTUCE PRODUCTION IN MINNEAPOLIS-SAINT PAUL, USA
We will consider a scenario of a major U.S. metropolitan area, Minneapolis-Saint Paul, becoming self-sufficient for one vegetable crop, lettuce. Lettuce production is the leading U.S. vegetable crop in terms of value, generating nearly $2 billion worth of product annually [77]. California and Arizona account for nearly all commercial lettuce production (both head and leaf lettuce) in the U.S. [77], but lettuce is also a common UA crop that is grown both indoors and outdoors. Minneapolis and Saint Paul have a combined population of 730,000 [78], and Americans consume an average of 2.63 kg of lettuce annually [77], resulting in an estimated annual lettuce demand of 1900 metric tons for Minneapolis and Saint Paul.

7.1 Importing lettuce from California

Because California accounts for over 70% of U.S. lettuce production, we will first consider the economic and environmental costs of importing lettuce from Monterey County, California. Lettuce yields on the California Central Coast exceed 46,000 kg/ha [79], resulting in a land requirement of 41.7 ha to meet the demand for Minneapolis-Saint Paul. The human labor requirement for this lettuce production comes to nearly 21,000 person-hours, mostly during harvest [79]. The estimated water requirement for irrigating the lettuce crop is 380,000 m^3 [80]. Including transportation, the total CO_2 emissions from lettuce production and distribution are approximately 1,770,000 kg [80] (Fig. 4). The total cost of production and distribution is approximately $1.2 million [79, 81].

7.2 Outdoor UA lettuce production

If the 1900 metric tons of lettuce were to be produced entirely through outdoor UA, it would require approximately 21 hectares of land, assuming two crops are produced annually (this calculation assumes that the entire year's lettuce supply would be produced during the ca. 5-month growing season, ignoring storage constraints). Minneapolis alone has over

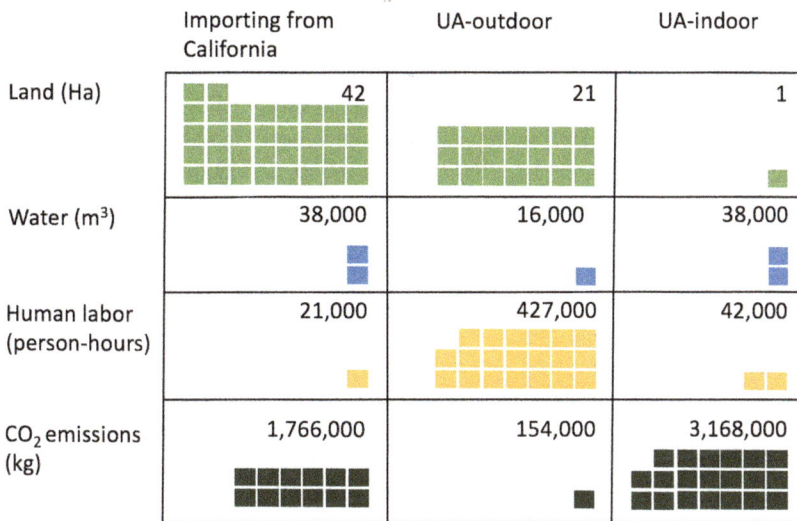

Figure 4: Estimated resources required, and pollution generate, in order to supply lettuce for the Twin Cities from importing from California, local outdoor UA production, and local indoor UA production.

315 hectares of vacant land [82], so the space requirement is not a limitation. This lettuce production would require approximately 16,000 m^3 of irrigation water to supplement average rainfall during the growing season [83]. This total volume of water represents less than 8% of the daily water consumption by Minneapolis [84], so it is not likely that water would be a limiting factor. Estimated total CO_2 emissions are 154,000 kg [80], less than 10% of the importation scenario (Fig. 4). Total human labor requirements, assuming an average of 2 person-hours/m^2 (conservatively assumed to be one-third lower than mean values for residential gardens reported in [43]) would exceed 427,300 person-hours. If this was paid labor and workers received $11.00/hour (current minimum wage in Minneapolis), total labor costs would exceed $4.7 million, several times higher than the total cost (production + transportation) of importing lettuce from California. The high turnover of commercial urban farms in Minneapolis and Saint Paul attests to the difficult economic barriers to small-scale urban production. As shown in a recent analysis for Sydney [43], human labor would likely be the limiting factor constraining the scaling up of lettuce production through outdoor UA, combined with the lack of production during the winter months.

7.3 Indoor UA lettuce production (Vertical Farm)

If the 1900 metric tons of lettuce were to be produced entirely through indoor UA, year-round lettuce demand could be supplied on only 1 hectare [85]. Total water demand for this crop would be approximately 38,000 m^3 [86]. Energy demand for lighting and heating would be approximately 350 kW, and heating would require approximately 388,000 therms [87]. Based on current energy infrastructure of the local utility, this energy use would generate 3,168,000 kg CO_2/year, nearly double that of the importation scenario (Fig. 4), at an annual utilities cost of $580,000 [87]. As the utility shifts to more renewable energy production (coal currently accounts for 40% of electricity production in Minnesota), CO_2 production would decrease.

Human labor requirements would depend on the degree of automation, but we estimate a requirement of 42,000 person-hours/year based on one local indoor production facility [88].

The capital costs of indoor production can be extremely high, requiring products to be sold at premium prices to restaurants and high-end grocery stores. Scaling up local lettuce production would quickly saturate the market for premium lettuce, constraining revenue for this industry. Illustrative of the economic challenges of indoor production, a large (0.8 ha) commercial aquaponics facility in Saint Paul that has previously received large local government subsidies, recently lost the support of its corporate backer due to not meeting business expectations and is in the process of shutting down [88]. Indoor production of lettuce to meet the demand of Minneapolis-Saint Paul is certainly feasible in terms of space and water requirements, but at a high energy and financial cost.

A supply chain disruption such as a severe drought in California, or a steep rise in petroleum prices that would raise transportation costs, could make UA production (both indoor and outdoor) more economically favorable. However, under current economic conditions, both indoor and outdoor UA struggle to compete with the relatively low production costs of large-scale commercial lettuce production. Aside from economic challenges, it is not clear that local production of lettuce is inherently more environmentally sustainable than importation under current conditions (Fig. 4). This analysis illustrates the importance of considering multiple metrics in assessing sustainability of UA.

8 CONCLUSION

Would a sustainable city be self-sufficient in food production? Modernist visions of sustainable cities suggest that this is both possible and desirable, but neither of these claims is self-evident. Self-sufficiency for at least some fruit and vegetable crops may be possible, but would likely require strong governmental social or economic incentives, such as in the U.S. Victory Garden program or Cuba's UA initiatives. However, even if self-sufficiency for certain crops is possible, it may not be desirable. Having a resilient food system requires redundancies to buffer against disruptions due to weather, pests, or other factors, and importing food from a variety of locations is one way to do that.

Meeting caloric and protein requirements of urban residents through UA is a much more daunting task, and the bioenergetic and economic feasibility has yet to be demonstrated. The production of animal protein (e.g. fish and chicken) through UA is likely to rely on imported animal feed, or if produced locally, would be competing for resources with food for direct human consumption. Shifting the bulk of food production for urban residents from rural areas to cities would require a dramatic change in the economic playing field, such as a drastic rise in fossil fuel costs that makes long-distance transportation economically prohibitive. Even then, the space, water, and energy requirements of agricultural production in cities would be in direct competition for other human uses, which would further drive up the cost of urban food production.

If self-sufficiency is not the objective of UA, then what is the optimal extent of UA in a sustainable city? The conceptual framework we have presented in this paper is meant to be a guide in exploring this question. Sustainability should necessitate using resources that are abundant while conserving limiting resources. The context of resource availability will differ from one city to another; and therefore, the metrics by which the sustainability of UA should be assessed will necessarily be context-dependent. For some cities, optimization of water use may be the driving factor, while for others, UA may capitalize on abundant ambient precipitation. UA in some cities with high labor costs may require optimization of human labor inputs, while in other cities with abundant human labor, UA may be a good source of jobs. The high value of real estate in some cities may drive UA to optimize spatial footprints in some cities, whereas in some post-industrial cities, the goal of UA may be to repurpose vacant land. Economic factors will play a central role in shaping the types of UA that develop in a given city, although some of the social benefits, as well as the cost of pollution, may not be accounted for in a free-market economy. Thus, local governments, non-profits, and other organizations have an important role to play in balancing the societal and environmental benefits and costs of UA.

The ecologist Eugene Odum referred to cities as 'parasites on the landscape' [89] since they rely on imports of food from, and the exportation of pollution to the hinterlands. Odum went on to note that parasites do not live long if they kill or damage their host. Moving cities towards sustainability is critical for the future of humanity, but the extent to which this requires self-sufficiency at the local scale is an open question. The extent to which cities can, or should, be self-sufficient in food production, is a simple question with a complex answer, we argue, with the answer depending on the socio-environmental context of each city. The role of urban agriculture in a sustainable city requires critical analysis, and the framework presented here offers one approach to this task.

ACKNOWLEDGEMENTS

GES thanks colleagues at the University of Minnesota Institute on the Environment for providing feedback on these ideas. This study was supported in part by a National Science Foundation CAREER award (award number 1651361) to GES. RM thanks his PhD advisors, Romina Rader and Paul Kristiansen, under whose supervision work that forms part of this paper was carried out, and acknowledges the funding provided in the form of an Australian Postgraduate Award and University of New England completion scholarship.

REFERENCES

[1] Burger, J.R., Allen, C.D., Brown, J.H., Burnside, W.R., Davidson, A.D., Fristoe, T.S., Hamilton, M.J., Mercado-Silva, N., Nekola, J.C., Okie, J.G. & Zuo, W., The macroecology of sustainability. *PLoS Biol*, **10(6)**, p. e1001345, 2012. https://doi.org/10.1371/journal.pbio.1001345

[2] Rees, W. & Wackernagle, M., Urban ecological footprints: Why cities cannot be sustainable–and why they are a key to sustainability. *Environmental Impact Assessment Review*, **16(4–6)**, pp. 223–248, 1996. https://doi.org/10.1016/s0195-9255(96)00022-4

[3] Weber, C.L. & Matthews, H.S., Quantifying the global and distributional aspects of American household carbon footprint. *Ecological Economics*, **66(2–3)**, pp. 379–391, 2008. https://doi.org/10.1016/j.ecolecon.2007.09.021

[4] Rothwell, A., Ridoutt, B., Page, G. & Bellotti, W., Environmental performance of local food: trade-offs and implications for climate resilience in a developed city. *Journal of Cleaner Production*, **114**, pp. 420–430, 2016. https://doi.org/10.1016/j.jclepro.2015.04.096

[5] Born, B. & Purcell, M., Avoiding the local trap: Scale and food systems in planning research. *Journal of Planning Education and Research*, **26(2)**, pp. 195–207, 2006. https://doi.org/10.1177/0739456x06291389

[6] Nogeire-McRae, T., Ryan, E.P., Jablonski, B.B.R., Carolan, M., Arathi, H.S., Brown, C.S., Saki, H.H., McKeen, S., Lapansky, E. & Schipanski, M.E., The role of urban agriculture in a secure, healthy, and sustainable food system. *BioScience*, **68**, pp. 748–759, 2018. https://doi.org/10.1093/biosci/biy071

[7] Harrison, J.L., *Pesticide Drift and the Pursuit of Environmental Justice*, MIT Press, 2011.

[8] Goddard, M.A., Dougill, A.J. & Benton, T.G., Scaling up from gardens: Biodiversity conservation in urban environments. *Trends in Ecology and Evolution*, **25**, pp. 90–98, 2010. https://doi.org/10.1016/j.tree.2009.07.016

[9] Susca, T., Gaffin, S.R. & Dell'Osso, G.R., Positive effects of vegetation: urban heat island and green roofs. *Environmental Pollution*, **159(8–9)**, pp. 2119–2126, 2011. https://doi.org/10.1016/j.envpol.2011.03.007

[10] Armstrong, D., A survey of community gardens in upstate New York: Implications for health promotion and community development. *Health Place*, **6(4)**, pp. 319–327, 2009. https://doi.org/10.1016/s1353-8292(00)00013-7

[11] de Zeeuw, H.R., Veenhuizen, V. & Dubbeling, M., The role of urban agriculture in building resilient cities in developing countries. *Journal of Agricultural Science*, **149(S1)**, pp. 153–163, 2011. https://doi.org/10.1017/s0021859610001279

[12] Mincyte, D. & Dobernig, K., Urban farming in the North American metropolis: rethinking work and distance in alternative food networks. *Environment Planning A: Economy and Space*, **48**, pp. 1767–1786, 2016. https://doi.org/10.1177/0308518x16651444

[13] McClintock, N., Why farm the city? Theorizing urban agriculture through a lens of metabolic rift? *Cambridge Journal of Regions Economy and Society*, **3(2)**, pp. 191–207, 2010. https://doi.org/10.1093/cjres/rsq005

[14] Turner, B., Embodied connections: Sustainability, food systems and community gardens. *Local Environment*, **16(6)**, pp. 509–522, 2011. https://doi.org/10.1080/13549839.2011.569537

[15] Frumkin, H., Healthy places: Exploring the evidence. *American Journal of Public Health*, **93**, pp. 1451–1456, 2003. https://doi.org/10.2105/ajph.93.9.1451

[16] Goodman, W. & Minner, J., Will the urban agricultural revolution be vertical and soilless? A case study of controlled environment agriculture in New York City. *Land Use Policy*, **83**, pp. 160–173, 2019. https://doi.org/10.1016/j.landusepol.2018.12.038

[17] Turner, W., Nakamura, T. & Dinetti, M., Global urbanization and the separation of humans from nature. *BioScience*, **54(6)**, p. 585, 2004. https://doi.org/10.1641/0006-3568(2004)054[0585:guatso]2.0.co;2

[18] McCormack, L.A., Laska, M.N., Larson, N.I. & Story, M., Review of the nutritional implication of farmers' markets and community gardens: A call for evaluation and research efforts. *Journal of the American Dietetic Association*, **110(3)**, pp. 399–408, 2010. https://doi.org/10.1016/j.jada.2009.11.023

[19] Boeing, H., Bechthold, A., Bub, A., Ellinger, S., Haller, D., Kroke, A., Leschik-Bonnet, E., Muller, M.J., Oberritter, H., Shulze, M., Stehle, P. & Watzl, B., Critical review: Vegetables and fruit in the prevention of chronic diseases. *European Journal of Nutrition*, **51(6)**, pp. 637–663, 2012. https://doi.org/10.1007/s00394-012-0380-y

[20] Joye, Y., Architectural lessons from environmental psychology: The case of biophilic architecture. *Review of General Psychology*, **11(4)**, pp. 305–328, 2007. https://doi.org/10.1037/1089-2680.11.4.305

[21] Ulrich, R., Evidence-based health-care architecture. *Lancet*, **368(12)**, pp. S38–S39, 2006. https://doi.org/10.1016/s0140-6736(06)69921-2

[22] Dimitri, C., Olberholtzer, L. & Pressman, A., Urban agriculture: Connecting producers with consumers. *British Food Journal*, **118**, pp. 603–617, 2016. https://doi.org/10.1108/bfj-06-2015-0200

[23] Caldeyro-Stajano, M., Simplified hydroponics as an appropriate technology to implement food security in urban agriculture. *Practical Hydroponics Greenhouses*, **76**, pp. 1–6, 2004.

[24] Voicu, I. & Been, V., The effect of community gardens on neighboring property values. *Real Estate Economics*, **36(2)**, pp. 241–283, 2008. https://doi.org/10.1111/j.1540-6229.2008.00213.x

[25] Vitiello, D. & Wolf-Powers, L., Growing food to grow cities? The potential of agriculture for economic and community development in the urban United States. *Community Development Journal*, **49(4)**, pp. 508–523, 2014. https://doi.org/10.1093/cdj/bst087

[26] Cleveland, D.A. et al., The potential for urban household vegetable gardens to reduce greenhouse gas emissions. *Landscape and Urban Planning*, **157**, pp. 365–374, 2017. https://doi.org/10.1016/j.landurbplan.2016.07.008

[27] Al-Chalabi, M., Vertical farming: Skyscraper sustainability? *Sustainable Cities and Society*, **18**, pp. 74–77, 2015. https://doi.org/10.1016/j.scs.2015.06.003

[28] Harada, Y., Whitlow, T.H., Walter, M.T., Bassuk, N.L., Russell-Anelli, J. & Schindelbeck, R.R., Hydrology of the Brooklyn Grange, an urban rooftop farm. *Urban Ecosystems*, **21(4)**, pp. 673–689, 2018. https://doi.org/10.1007/s11252-018-0749-7

[29] Metson, G.S. & Bennett, E.M., Phosphorus cycling in Montreal's food and urban agricultural systems. *PLoS One*, **10(3)**, p. e0120726, 2015.

[30] Dewaelheyns, V., Elsen, A., Vandendriessche, H. & Gulinck, H., Garden management and soil fertility in Flemish domestic gardens. *Landscape and Urban Planning*, **116**, pp. 25–35, 2013. https://doi.org/10.1016/j.landurbplan.2013.03.010

[31] Small, G., Shrestha, P. & Kay, A., The fate of compost-derived phosphorus in urban gardens. *International Journal of Design & Nature and Ecodynamics*, **13(4)**, pp. 415–422, 2018. https://doi.org/10.2495/dne-v13-n4-415-422

[32] Lin, B.B., Philpot, S.M., & Jha, S., The future of urban agriculture and biodiversity-ecosystem services: Challenges and next steps. *Basic and Applied Ecology*, **16(3)**, pp. 189–201, 2015.

[33] Isaacs, R., Tuell, J., Fielder, A., Gardiner, M. M., & Landis, D., Maximising arthropod-mediated ecosystem services in agricultural landscapes: The role of native plants, *Frontiers in Ecology and the Environment*, **7**, pp. 196–203, 2009.

[34] Clinton, N., Stuhlmacher, M., Miles, A., Uludere Aragon, N., Wagner, M., Georgescu, M., Herwig, C. & Gong, P., A global geospatial ecosystem services estimate of urban agriculture. *Earth's Future*, **6**, pp. 40–60, 2018. https://doi.org/10.1002/2017ef000536

[35] Keeler, B.L., Hamel, P., McPhearson, T., Hamman, M.H., Donahue, M.L., Meza Prado, K.A., Arkema, K.K., Bratman, G.N., Brauman, K.A., Finlay, J.C., Guerry, A.D., Hobbie, S.E., Johnson, J.A., MacDonald, G.K., McDonald, R.I., Neverisky, N. & Wood, S.A., Social-ecological and technological factors moderate the value of urban nature, *Nature Sustainability*, **2(1)**, p. 29, 2019. https://doi.org/10.1038/s41893-018-0202-1

[36] Martelozzo, F., Landry, J.S., Plouffe, D., Seufert, V., Rowhani, P. & Ramankutty, N., Urban agriculture: A global analysis of the space constraint to meet urban vegetable demand. *Environmental Research Letters*, **9(6)**, p. 064025, 2014. https://doi.org/10.1088/1748-9326/9/6/064025

[37] Stanhill, G., An urban agro-ecosystem: The example of nineteenth-century Paris. *Agro-Ecosystems*, **3**, pp. 269–284, 1976. https://doi.org/10.1016/0304-3746(76)90130-x

[38] Nixon, P.A. & Ramaswami, A., Assessing current local capacity for agrifood production to meet household demand: Analyzing select food commodities across 377 U.S. Metropolitan Areas. *Environmental Science and Technology*, **52(18)**, pp. 10511–10521, 2018. https://doi.org/10.1021/acs.est.7b06462

[39] Grewal, S.S. & Grewal, P.S., Can cities become self-reliant in food? *Cities*, **29(1)**, pp. 1–11, 2012. https://doi.org/10.1016/j.cities.2011.06.003

[40] Rodriguez, O., *London Rooftop Agriculture: A Preliminary Estimate of Productive Potential*, Master Thesis, Cardiff: Welsh School of Architecture, 2009.

[41] Caplow, T., Building integrated agriculture: philosophy and practice. In *Urban Futures 2030: Urban Development and Urban Lifestyles of the Future*, ed. Heinrich Böll Foundation, Heinrich-Böll-Stiftung, Berlin, Germany, pp. 54–58, 2009.

[42] Kong, A.Y.Y., Rosenzweig, C. & Arky, J., Nitrogen dynamics associated with organic and inorganic inputs to substrate commonly used on rooftop farms. *HortScience*, **50(6)**, pp. 806–813, 2015. https://doi.org/10.21273/hortsci.50.6.806

[43] McDougall, R., Kristiansen, P., & Rader, R., Small-scale urban agriculture results in high yields but requires judicious management of inputs to achieve sustainability. *Proceedings of the National Academy of Sciences USA*, **116(1)**, pp. 129–134, 2019. https://doi.org/10.1073/pnas.1809707115

[44] Despommier, D., The rise of vertical farms. *Scientific American*, **301(5)**, pp. 80–87, 2009. https://doi.org/10.1038/scientificamerican1109-80

[45] Artmann, M. & Sartison, K., The role of urban agriculture as a nature-based solution: A review for developing a systematic assessment framework. *Sustainability*, **10(6)**, p. 1937, 2018. https://doi.org/10.3390/su10061937

[46] Ramaswami, A., Boyer, D., Nagpure, A.S., Fang, A., Bogra, S., Bakshi, B., Cohen, E. & Rao-Ghorpade, A., An urban systems framework to assess the trans-boundary food-energy-water nexus: implementation in Delhi, India. *Environmental Research Letters*, **12(2)**, p. 025008, 2017. https://doi.org/10.1088/1748-9326/aa5556

[47] Dmitri, C., Oberholtzer, L. & Pressman, A., Urban agriculture: Connecting producers with consumers. *British Food Journal*, **118**, pp. 603–617, 2016. https://doi.org/10.1108/bfj-06-2015-0200

[48] Despommier, D., Farming up the city: the rise of urban vertical farms. *Trends in Biotechnology*, **31(7)**, pp. 388–389, 2013. https://doi.org/10.1016/j.tibtech.2013.03.008

[49] Cook, R.L. & Calvin, L., Greenhouse tomatoes change the dynamics of the North American fresh tomato industry. *U.S. Department of Agriculture Economic Research Report*, **2**, pp. 1–11, 2005.

[50] Astee, L.Y. & Kishnani, N., Building integrated agriculture: Utilising rooftops for sustainable food crop cultivation in Singapore. *Journal of Green Building*, **5(2)**, pp. 105–113, 2010. https://doi.org/10.3992/jgb.5.2.105

[51] Sauerborn, J., Skyfarming: An alternative to horizontal croplands. *Resource Magazine*, **18**, pp. 938–941, 2011.

[52] Ellingsen, E. & Despommier, D., The vertical farm: The origin of a 21st century architectural typology. *CTBUH Journal*, **3**, pp. 26–34, 2008.

[53] Delor, M., *Current State of Building-Integrated Agriculture, Its Energy Benefits and Comparisons with Green Roofs*, University of Sheffield, Sheffield, UK, 2011.

[54] Reinhardt, W., Albright, L. & de Villiers, D.S., Energy investments and CO_2 emissions for fresh produce imported into New York State compared to the same crops grown locally. *New York State Energy Research and Development Authority*, pp. 8–10, 2008.

[55] Specht, K., Siebert, R., Hartmann, I., Freisinger, U.B., Sawicka, M., Werner, A., Thomaier, S., Henckel, D., Walk, H. & Dierich, A., Urban agriculture of the future: An overview of sustainability aspects of food production in or on buildings. *Agriculture and Human Values*, **31(1)**, pp. 33–51, 2014. https://doi.org/10.1007/s10460-013-9448-4

[56] Bhanoo, S., Vertical farms will be big, but for whom? Indoor farming might help feed millions, or at least make millions, https://www.fastcompany.com/3039087/vertical-farms-will-be-big-but-for-who (accessed 7 May 2019).

[57] de Nijs, B., Does vertical farming make sense? *Hortidaily.com*, https://www.hortidaily.com/article/35974/Does-vertical-farming-make-sense/ (accessed 7 May 2019).

[58] Vogl, C.R., Axmann, P. & Vogl-LuKasser, B., Urban organic farming in Austria with the concept of Selbstemte ('self-harvest'): An agronomic and socio-economic analysis. *Renewable Agriculture and Food Systems*, **19(2)**, pp. 67–79, 2004. https://doi.org/10.1079/rafs200062

[59] Ackerman, K., Dahlgren, E. & Xue, X., *Sustainable Urban Agriculture: Confirming Viable Scenarios for Production*, New York State Energy Research and Development Authority: Albany, New York, 2013.

[60] Taylor, J.R. & Lovell, S.T., Urban home food gardens in the Global North: research traditions and future directions. *Agriculture and Human Values*, **31(2)**, pp. 85–305, 2014. https://doi.org/10.1007/s10460-013-9475-1

[61] Swaney, D.P., Howarth, R.W. & Hong, B., Nitrogen use efficiency and crop production: Patterns of regional variation in the United States, 1987–2012, *Science of the Total Environment*, **635**, pp. 498–511, 2018. https://doi.org/10.1016/j.scitotenv.2018.04.027

[62] Suh, S. & Yee, S., Phosphorus use-efficiency of agriculture and food system in the U.S. *Chemosphere*, **84(6)**, pp. 806–813, 2011. https://doi.org/10.1016/j.chemosphere.2011.01.051

[63] Hoornweg, D. & Bhada-Tata, P., *What a Waste - A Global Review of Solid Waste Management*, World Bank Urban Development Series Knowledge Papers, **15**, 2012.

[64] Fletcher, T.D., Deletic, A., Mitchell, V.G. & Hatt, B.E., Reuse of urban runoff in Australia: A review of recent advances and remaining challenges. *Journal of Environmental Quality*, **37(5_Supplement)**, pp. S–116, 2008. https://doi.org/10.2134/jeq2007.0411

[65] Barthel, S., Parker, J. & Ernstson, H., Food and green space in cities: A resilience lens on gardens and urban environmental movements. *Urban Studies*, **52(7)**, pp. 1321–1338, 2015. https://doi.org/10.1177/0042098012472744

[66] Barthel, S., Folke, C. & Colding, J., Social-ecological memory in urban gardens-Retaining the capacity for management of ecosystem services. *Global Environmental Change*, **20(2)**, pp. 255–265, 2010. https://doi.org/10.1016/j.gloenvcha.2010.01.001

[67] Endres, A.B. & Endres, J.M., Homeland security planning: What victory gardens and Fidel Castro can teach us in preparing for food crises in the United States. *Food and Drug Law Journal*, **64**, pp. 405–439, 2009.

[68] Warwick, H., Cuba's organic revolution. *Forum for Applied Research and Public Policy*, **16**, pp. 54–58, 2001.

[69] Brown, M.E., Antle, J.M., Backlund, P., Carr, E.R., Easterling, W.E., Walsh, M.K., Ammann, C., Attavanich, W., Barrett, C.B., Bellemare, M.F., Dancheck, V., Funk, C., Grace, K., Ingram, J.S.I., Jiang, H., Maletta, H., Mata, T., Murray, A., Ngugi, M., Ojima, D., O'Neill, B. & Tebaldi, C., *Climate Change, Global Food Security, and the U.S. food system*. U.S. Department of Agriculture, 2015.

[70] Ostrum, E., Polycentric systems for coping with collective action and global environmental change. *Global Environmental Change*, **20(4)**, pp. 550–557, 2010. https://doi.org/10.1016/j.gloenvcha.2010.07.004

[71] Qui, G., Li, H., Zhang, Q., Chen, W., Liang, X. & Li, X., Effects of evapotranspiration on mitigation of urban temperature by vegetation and urban agriculture. *Journal of Integrative Agriculture*, **12(8)**, pp. 1307–1315, 2013. https://doi.org/10.1016/s2095-3119(13)60543-2

[72] Egerer M.H., Lin, B.B., Threlfall, C.G. & Kendal, D., Temperature variability influences urban garden plant richness and gardener water use behavior, but not planting decisions. *Science of the Total Environment*, **646**, pp. 111–120, 2019. https://doi.org/10.1016/j. scitotenv.2018.07.270

[73] Barthel, S. & Isendahl, C., Urban gardens, agriculture, and water management: Sources of resilience for long-term food security in cities. *Ecological Economics,* **86**, pp. 224–234, 2013. https://doi.org/10.1016/j.ecolecon.2012.06.018

[74] Somerville, C. & Ferrand, C., Aquaponics in Gaza, *Field Exchange 46: Special Focus on Urban Food Security & Nutrition*, September 2013.

[75] Foskett, D., *Food Security and Small-Scale Aquaponics: A Case Study on the Northern Mariana Island of Rota*, M.A. Thesis, University of Oregon, 2014.

[76] Moore, A., This local agtech startup wants you to forget everything you know about farming. *Upstate Business Journal*, Online: https://upstatebusinessjournal.com/this-local-agtech-startup-wants-you-to-forget-everything-you-know-about-farming/, 5 March 2018 (accessed 7 May 2019).

[77] Agricultural Marketing Research Center, Lettuce. Online: https://www.agmrc.org/commodities-products/vegetables/lettuce (accessed 7 May 2019).

[78] Pioneer Press, Census: Minneapolis-St. Paul metro adds more than 250,000 residents since 2010. Online https://www.twincities.com/2018/03/22/census-minneapolis-st-paul-metro-adds-more-than-250000-residents-since-2010/ (accessed 15 May 2019).

[79] University of California Cooperative Extension, Sample production costs for wrapped iceberg lettuce sprinkler irrigated–40-inch beds: Central Coast, 2010. Online: https://coststudyfiles.ucdavis.edu//uploads/cs_public/a4/bb/a4bb20f0-4bfe-404e-b47e-b7a634ca80b5/2010lettuce_wrap_cc.pdf (accessed 15 May 2019).

[80] Goldstein, B.P., Hauschild, M.Z., Fernandez, J. & Birkved, M., Testing the environmental performance of urban agriculture as a food supply in northern climates. *Journal of Cleaner Production*, **135**, pp. 984–994, 2016. https://doi.org/10.1016/j. jclepro.2016.07.004

[81] Ronan, D., Cost of operating a truck up 6% to $1.69 per mile, ATRI report says, *Transpor Topics*, Online: https://www.ttnews.com/articles/cost-operating-truck-6-169-mile-atri-report-says (accessed 15 May 2019).

[82] City of Minneapolis Land Capacity Analysis, 2010. Online: http://www.minneapolismn.gov/www/groups/public/@cped/documents/webcontent/convert_261135.pdf, (accessed 15 May 2019).

[83] U.S. Climate Data. Online: https://www.usclimatedata.com/climate/minneapolis/minnesota/united-states/usmn0503, (accessed 15 May 2019).

[84] Minneapolis Water Treatment and Distribution Services. Online: http://www.minneapolismn.gov/publicworks/water/water_waterfacts (accessed 15 May 2019).

[85] Kubota, C., Controlled Environment Agriculture for Urban Food Production, 2018 Urban Food Systems Symposium, Minneapolis, MN, 2018.

[86] Barbosa, G.L., Gadelha, F.D.A., Kublick, N., Proctor, A., Reichelm, L., Weissinger, E., Wohlleb, G.M. & Halden, R.U., Comparison of land, water, and energy requirements of lettuce grown using hydroponics vs. conventional agricultural methods. *International Journal of Environmental Research and Public Health*, **12(6)**, pp. 6879–6891, 2015. https://doi.org/10.3390/ijerph120606879

[87] Xcel Energy, Carbon dioxide emission intensities, 2018. Online: https://www.xcelenergy.com/staticfiles/xe-responsive/Environment/Carbon/Xcel-Energy-Carbon-Dioxide-Emission-Intensities.pdf (accessed 15 May 2019).

[88] Riley, M., Pentair shutting down urban organics aquaponics facility in St. Paul. *Minneapolis/St. Paul Business Journal*, 15 May 2019, Online: https://www.bizjournals.com/twincities/news/2019/05/15/pentair-shutting-down-urban-organics-aquaponics.html (accessed 20 May 19).

[89] Odum, E. P., *Ecology and Our Endangered Life-Support Systems*, Sunderland, MA: Sinauer Associates, 1989.

[90] CoDrye, M., Fraser, E.D.G. & Landman, K., How does your garden grow? An empirical evaluation of the costs and potential of urban gardening. *Urban Forestry & Urban Greening*, **14(1)**, pp. 72–79, 2015. https://doi.org/10.1016/j.ufug.2014.11.001

[91] Haberman, D., Gillies, L., Canter, A., Rinner, V., Pancrazi, L. & Martellozzo, F., The potential of urban agriculture in Montreal: A quantitative assessment. *ISPRS International Journal of GeoInformation*, **3(3)**, pp. 1101–1117, 2014. https://doi.org/10.3390/ijgi3031101

[92] Hara, Y., Murakami, A., Tsuchiya, K., Palijon, A.M. & Yokohari, M., A quantitative assessment of vegetable farming on vacant lots in an urban fringe area in Metro Manila: Can it sustain long-term vegetable demand? *Applied Geography*, **41**, pp. 195–206, 2013. https://doi.org/10.1016/j.apgeog.2013.04.003

[93] MacRae, R., Gallant, E., Patel, S., Michalak, M., Bunch, M. & Schaffner, S., Could Toronto provide 10% of its fresh vegetable requirements from within its own boundaries? Matching consumption requirements with growing spaces, *Journal of Agriculture, Food Systems, and Community Development*, **1(2)**, pp. 105–127, 2010. https://doi.org/10.5304/jafscd.2010.012.008

[94] Orsini, F., Gasperi, D., Marchetti, L., Piovene, C., Draghetti, S., Ramazzotti, S., Bazzocchi, G. & Gianquinto, G., Exploring the production capacity of rooftop gardens (RTGs) in urban agriculture: the potential impact on food and nutrition security, biodiversity and other ecosystem services in the city of Bologna. *Food Security,* **6(6)**, pp. 781–792, 2014. https://doi.org/10.1007/s12571-014-0389-6

[95] Johnson, M.S., Lathuillière, M.J., Tooke, T.R. & Coops, N.C., Attenuation of urban agricultural production potential and crop water footprint due to shading from buildings and trees. *Environmental Research Letters*, **10(6)**, pp. 1–11, 2015. https://doi.org/10.1088/1748-9326/10/6/064007

[96] Lee, G.G., Lee, H.W. & Lee, J.H., Greenhouse gas emission reduction effect in the transportation sector by urban agriculture in Seoul, Korea. *Landscape and Urban Planning,* **140**, pp. 1–7, 2015. https://doi.org/10.1016/j.landurbplan.2015.03.012

[97] Hara, Y., McPhearson, Sampei, T. & McGrath, B., Assessing urban agriculture potential: A comparative study of Osaka, Japan and New York City, United States, *Sustainability Science*, **13(4)**, pp. 937–952, 2018. https://doi.org/10.1007/s11625-018-0535-8

[98] Sioen, G.B., Sekiyama, M., Terada, T. & Yokohari, M., Post-disaster food and nutrition from urban agriculture: A self-sufficiency analysis of Nerima ward, Tokyo. *International Journal of Environmental Research and Public Health*, **14(7)**, p. 748, 2017. https://doi.org/10.3390/ijerph14070748

[99] Clark, K.H. & Nicholas, K.A., Introducing urban food forestry: A multifunctional approach to increase food security and provide ecosystem services. *Landscape Ecology*, **28(9)**, pp. 1649–1669, 2013. https://doi.org/10.1007/s10980-013-9903-z

[100] Colosanti, K. & Hamm, M.W., Assessing the local food supply capacity of Detroit, Michigan. *Journal of Agriculture, Food Systems, and Community Development*, **1(2)**, pp. 41–58, 2010. https://doi.org/10.5304/jafscd.2010.012.002

CONCLUDING REMARKS FROM THE IMPLEMENTATION OF SMART LOW-ENERGY DISTRICTS IN THE GROWSMARTER PROJECT

ALAIA SOLA, MANEL SANMARTI & CRISTINA CORCHERO
Energy Systems Analytics Research Group, IREC Catalonia Institute for Energy Research, Spain

ABSTRACT

As large consumers of energy, cities offer the opportunity for significant energy savings in relation to the implementation of energy-efficiency measures. In this context, the cities of Barcelona, Cologne and Stockholm, together with a diverse group of stakeholders from public and private sectors, joined to create the GrowSmarter project. GrowSmarter seeks to demonstrate and stimulate the uptake of Smart Solutions in energy, infrastructure and transport, to provide other cities with insights and create a ready market to support the transition to a sustainable Europe. With the objective of promoting and developing low-energy districts, a set of solutions were tested aiming to reduce their environmental impact. These are classified in three blocks: building energy retrofitting, energy consumption visualization platforms and local energy generation with smart management. All these actions have been technically and economically evaluated in GrowSmarter, and the results are presented in this article. The project has analysed different impacts of active and passive retrofitting measures in building energy performances and the feasibility of the proposed business models behind them. Energy visualization platforms have proven to be a promising tool to engage end users, but there is still work to do to define successful business models. The assessment of the deployment of local energy generation units shows that the corresponding regulation differs to a significant extent among countries. A clear and harmonized regulation according to the current state of technology is required in order to fully deploy distributed energy resources at commercial level. Finally, besides guaranteeing the correct implementation and operation of energy-efficiency measures, communication and information campaigns are key to build trust and ensure user acceptance. Working on building users' awareness and acceptance has proven to be a must in order to succeed in making low-energy districts the preferred path in urban development.
Keywords: smart city, energy efficiency, energy retrofitting, HEMS, local energy, low-energy districts, H2020.

1 INTRODUCTION

Approximately 70% of the world population will live in urban areas by 2050, according to UN predictions [1]. This means that cities are very likely to be one of the largest groups of energy consumers in the world and, therefore, potentially one of the biggest emitters of greenhouse gases. To address this challenging forecast, municipal governments around the world are developing ambitious programs for long-term emission reduction. This implies a change in the energy model of cities towards a more sustainable one, which is one of the main characteristics behind the concept of 'smart city'.

Across Europe, cities are adopting smart and sustainable development programs. In order to promote this development, the European Commission launched in 2012 the European innovation partnership on smart cities and communities [2], which joins together European cities, industry leaders and representatives of civil society to make urban areas smarter. In addition, there are various European funding programs and instruments that support the transition to smart cities, such as the Horizon 2020 and COSME programs [3,4]. In this way, the European Union is promoting the demonstration of the technical and economic impact of various solutions towards the smart and sustainable development of cities.

Developing low-energy districts is one of the main action areas to be addressed in order to transform cities towards more sustainable ones. In fact, the European construction sector faces a major challenge to reduce the emissions by almost 90% in 2050. This requires new innovative solutions and services to be rapidly implemented in the market, such as affordable and sustainable building energy retrofit solutions at a large scale.

2 THE GROWSMARTER PROJECT

The Municipalities of Barcelona, Cologne and Stockholm, together with a diverse group of partners from public and private sectors, have been leading since 2015 the GrowSmarter project [5] funded by the Horizon 2020 program of the European Commission. Horizon 2020 is a program designed to support R&D activities and aimed at researchers, companies, technology centres and public entities. Within this program, GrowSmarter belongs to a specific call named 'smart and sustainable cities' [6] (SCC1), which aims to promote demonstrative or 'lighthouse' projects. The main objective of this type of project is to connect cities, industry and citizens in order to demonstrate in real life a variety of technical solutions and business models to build smarter cities, which are scalable and replicable. Towards this aim, GrowSmarter has implemented and demonstrated the viability of '12 smart city solutions' in the fields of energy, infrastructure and transport.

2.1 Low-energy districts in GrowSmarter

Identified as one of the instruments to improve the quality of life of citizens, energy efficiency in buildings has become one of the main features of smart cities. This article focuses on the outcome of the work performed within the GrowSmarter's energy work package called 'low-energy districts', whose main objective is the deployment of energy-efficiency measures to reduce the environmental impact of the existing building stock in cities.

In total, 123,000 m² of constructed surface area have been retrofitted among the three lighthouse cities, involving private and public buildings, as well as commercial and residential buildings. Modern houses are highly energy efficient, but one-third of Europe's housing stock was built between 1950 and 1970 when the technologies and materials used were much less efficient. By refurbishing these older buildings using new construction techniques and installing new efficient equipment, the amount of energy they use can be substantially reduced.

The project has also promoted the deployment of Home Energy Management Systems (HEMS) in the three cities to raise awareness about responsible energy consumption among citizens and give tools for appliances' automation. Providing information on real-time energy usage to tenants is a key tool to help them see and reduce their own environmental footprint. Moreover, in Barcelona and Stockholm, energy surveillance platforms have also been installed in commercial buildings, with the aim of simplifying their operation and maintenance and identifying potential energy-efficiency improvements.

Finally, the integration of local energy generation in buildings has been demonstrated in the three cities with different approaches: on-site renewable electricity generation with photovoltaics (PV) with batteries under smart control, connection of buildings to district heating and cooling (DHC) networks and combination of PV, heat pumps and DHC together with storage under smart control.

By deploying the aforementioned solutions, all three lighthouse cities demonstrate the feasibility and real impact of a set of available innovative technologies. This contributes to the transformation of the existing city building stock towards low-energy districts.

3 LESSONS LEARNED AND RECOMMENDATIONS

This section presents the lessons learned and recommendations gathered by the private and public stakeholders during the 3 years of implementation. To date, the GrowSmarter project is facing its last stage, where all measures are completely executed and the monitoring phase has been running for more than 1 year. This will allow to draw future results on the technical and economic performance of each technology in the three cities.

3.1 Lessons learned and recommendations to industrial partners

3.1.1 Building energy retrofitting

In GrowSmarter, industrial partners such as utilities and project development and construction groups have implemented retrofitting actions to upgrade the energy performance of existing residential and commercial buildings. The casuistry in each of the three lighthouse cities is different in terms of municipal regulations, climate and the proposed business models behind an energy retrofitting project. However, in general, energy audits during the design stage are fundamental to adjust the technical solutions to each specific building. It is also recommended to evaluate the possible sources of energy in the surroundings of buildings or facilities to look for waste heat integration solutions that can benefit multiple actors. In the case of Stockholm, the energy assessment of surroundings allowed recovering residual heat from a data centre to heat up a commercial building retrofitted by the Municipality.

In mild climates, such as the Mediterranean, energy savings from passive energy retrofitting (i.e. improvement of building thermal envelope) have proven very long payback time compared to similar actions in cold climates. If public subsidies or funding are not available, adding active solutions (e.g. replacement of old Heating, Ventilating and Air Conditioning (HVAC) and lighting equipment by new equipment with higher energy efficiency) to the passive measures has evidenced to be an alternative to shorten the payback time. In those cases, the monetization of the revaluation of the property after energy retrofitting is also crucial to help reaching the economic feasibility of passive measures. Related to commercial buildings, the project has tested in Barcelona a business model in the form of a private Energy Services Company (ESCo) participation in already approved structural refurbishment projects. This approach has shown to be a good opportunity to include energy-efficiency measures that would otherwise be omitted in commercial buildings renovation plans. The collaboration allows sharing costs between the constructor and the ESCo during the renovation works, besides adding a stronger commitment to guarantee a predefined energy savings target. In Cologne and Stockholm, large retrofitting projects have been implemented in residential settlements which involved the retrofitting of several buildings owned by the public administration. In these cases, avoiding multiple ownership in the residential sector has shown to facilitate the implementation. In order to promote the retrofitting of multiownership residential buildings, it is important to explain the added value of implementing energy-efficiency solutions compared to simple building retrofitting. The message on energy retrofitting value has to be broadened by a variety of arguments: (1) saving money, (2) better indoor climate, (3) getting more control of your energy use, (4) reducing the environmental impact, (5) increasing the asset value.

3.1.2 Home automation tools and energy surveillance platforms

The market of smart services at home has also been explored in the three lighthouse cities through the installation of HEMS prototypes in the project. These tools are able to offer different services other than real-time energy consumption visualization (e.g. remote control of appliances through smart plugs). However, in order to offer energy bill reduction to clients, the access to granular energy consumption data is crucial as well as the increase of users' awareness. Therefore, both qualitative data and an active user are required. Teaching the tenants how to use the smart home solution (and the associated costs of the information campaigns) is an essential and sometimes obviated activity to guarantee the success of HEMS. In fact, one of the barriers encountered in the project has been social acceptance towards this type of tools. Data protection laws must be complied, and therefore each user/client must give permission for the treatment of his or her granular energy consumption data. For example, the prototype of the smart home system offered in Cologne could not be widely deployed because the targeted tenants were not interested in this kind of technology and did not give their consent to install uncertified electricity meters to obtain detailed data. This limited to a large extent the potential functionalities of the tool. An intensive market study to investigate the interests of target residents is crucial, which in turn must be accompanied by engagement campaigns to raise interest among potential clients. In reference to the state of technology, HEMS deployed in the project have shown some technical complexity related to maintenance, data communication and technology obsolescence. Notwithstanding, it is expected that the open home net will become a more common infrastructure in apartments, offering shared sensors and actuators. By sharing the sensors needed to provide different services in apartments, it will be possible to add many services to a lower cost and facilitate the upscaling of smart home systems. Depending on the price of electricity in each country, reaching a break-even between the potential electricity savings offered by the HEMS and the price that clients have to pay for the service may be hard to achieve. Therefore, in order to enhance the economic feasibility of energy consumption visualization platforms, it is advisable to find synergies with other home services and look for opportunities to sell the platforms with integrated services packages.

3.1.3 Local energy generation

In the framework of promoting distributed local energy generation in the city, the project has tested the integration of advanced control systems to benefit from intelligent management of PV, heat pumps and battery systems based on weather and consumption forecasts, among others. Smart control of distributed electricity generation is a technology under development these days, and the experience in GrowSmarter has shown that the variability of communication protocols among battery and inverter manufacturers is a barrier to the technical feasibility of smart external control. Controlling the devices as planned by the energy management system definition has proven technically feasible, but still challenging. The current state of technology does not always guarantee that the equipment will respond to the external commands as expected.

Smart control of local energy generation units also allows the characterization of how the electrical power is used over time in both residential and commercial buildings. This information can be used to search for new businesses and offer services of power equalization among different types of buildings. Finally, smart management has also been applied to existing district heating (DH) networks in Stockholm, where DH technology is very well-established. In this case, the economic feasibility and positive environmental impact of recovering waste heat from a data centre and a supermarket into the DH network has been proven. This innovative business models allows the heat suppliers to avoid cooling system costs and get benefits instead.

3.2 Lessons learned and recommendations to local governments

3.2.1 Social acceptance
Local governments are important players to reduce social barriers for the implementation of solutions towards low-energy districts. Public administration should not only be a source of incentives or promotion of energy-efficiency solutions for the city building stock but also of campaigns for citizens' awareness and engagement in energy-efficiency matters. Social acceptance has been found as one of the main challenges in GrowSmarter. The implementation of energy-efficiency measures in buildings has a direct impact in citizens' quality of life.

3.2.2 Regulation
As observed in Growsmarter, the impact of local regulation can play a significant role to both foster or suppress building energy retrofitting upscaling. Examples of good practices are the enforcement of building codes and the promotion of the connection of buildings to efficient district energy networks in the areas of the city where the infrastructure is available. In Barcelona, for example, an ordinance requires most of the new and retrofitted buildings to include solar PV on their rooftops, having a positive impact on the path to low-energy districts.

In large-scale home renovation plans which involve multiple building owners, the role of the administration can have a very positive impact as a trustful actor that encourages citizens to get on-board. Campaigns promoted by the Public administration generate more trust among citizens in front of privately promoted campaigns. In this sense, involving the public sector (i.e. city municipalities) with a leadership role in large-scale building energy retrofitting projects is seen as a promising tool to upscale energy retrofitting among multi-ownership buildings. The leadership role by the administration may involve functions such as managing the grants, proposing the most appropriate financing for the owners, managing the payments and delays, etc.

3.2.3 Public housing
In the three lighthouse cities, public housing has been retrofitted to reach energy savings at different levels. Public housing owners are key institutional stakeholders to be mobilized to reduce the energy consumption of the residential sector. As long-term managers of their housing stock, public housing owners have to anticipate upcoming regulations on (existing) buildings in order to avoid any extra costs of future refurbishments. The decision-making capacity and technical expertise in this sector are high.

Tenants of public housing have to be kept informed right from the very beginning of the process to foster acceptance and avoid the "rebound effect" (offset of the beneficial effects of the energy retrofitting due to behavioural responses) and effectively achieve energy savings. It is recommended to pursue the creation of a sense of ownership and understanding among tenants.

3.2.4 Heritage buildings
In Stockholm and Barcelona, the energy retrofitting actions have also targeted heritage buildings which have become public commercial buildings. These projects have shown that municipal regulations can enable the implementation of energy-efficiency criteria and on-site renewable energy generation in heritage buildings, instead of restricting their technical feasibility with protective regulations. In general, it has been observed that there is a lack

of regulatory initiatives in the lighthouse cities for reconciling energy efficiency/renewable energy with heritage conservation concerns, which could broaden the possibilities for the modernization of listed buildings in cities. It is important to avoid too many exceptions in the energy savings obligations of building renovation, because making heritage buildings sustainable is just as important as preserving their history. Using public municipal buildings as showcases for low-energy building design is seen as a useful tool to encourage private actors to invest in building energy performance upgrade and citizen engagement in general.

3.3 Lessons learned and recommendations to regional and national governments

3.3.1 District-scale energy renovation plans

Municipalities cannot face alone the promotion of district-scale energy retrofitting projects through investments or subsidies. Notwithstanding, regional and national incentives can bring an opportunity for municipalities to promote the decrease of the environmental impact of their building stock. In order to fairly quantify public subsidies and investments in the development of regional and national strategies, health benefits and the associated cost savings for the public healthcare system must be accounted. In fact, it is essential to assess and include all the externalities of energy retrofitting projects, i.e. social benefits in the form of non-economic benefits. District-scale energy renovation plans have several social benefits for citizens other than economic savings, e.g. increase of buildings' value or the prevention of young people to leave their neighbourhood.

It is recommended that subsidies for the residential sector are quantified considering the potential energy savings calculated based on real demands. The deployment of retrofitting actions in Barcelona has shown that the heating demands of residential buildings are often overestimated in mild climates due to the absence of heating system use by dwellers. Theoretical ratios do not represent the real energy consumption of a large share of the dwellings and, therefore, the impact of energy retrofitting works is sometimes lower than expected.

3.3.2 Local energy generation in urban environment

The upscaling and economic feasibility of local energy generation is highly dependent on national regulation. The smart management of on-site solar energy generation, heat pumps and/or battery storage demonstrated in GrowSmarter offers a new approach to integrate buildings in local energy communities, thus exploiting building flexibility and maximizing the investment in solar power. The associated benefits for the end user will increase when national regulation includes demand–response aspects, and this in turn will boost the scalability potential of smart home service. Using the tool for demand–response at the overall building level (and not at the apartment level) is seen as a promising service to increase the tool's replication potential. In fact, the current lack of flexibility to trade with energy and the sometimes unpredictable changes of laws build a state of no legal security for the scalability of local renewable electricity generation in cities, which claims for a political change. Regulation in electricity self-consumption (and electricity markets in general) differs in European countries. A regulation-free zone to prove hypotheses for local energy communities in the urban environment would contribute in the more restrictive countries to demonstrate the technical and economic feasibility. In a similar way, only if regulations are harmonized and updated according to the current state of technology, the exploitation of data from smart meters can be fully deployed. In practice, it is still not possible to consider a single model throughout the European Union that allows the scalability of products and services associated to the use of detailed electricity consumption data from smart meters (dependent on the regulatory frame and the available IT infrastructure in each country). A non-discriminatory access to data from smart-metering systems would avoid duplication of devices for the detailed monitoring of electricity consumption at dwelling level.

4 CONCLUSIONS

The deployment in the three lighthouse cities of a variety of energy-efficient solutions aiming at lowering the environmental impact of the districts has led to several learnings and recommendations for the future.

It is well known that, in general terms, a modernization of the existing building stock in European countries is required. In this line, it is expected that building retrofitting strategies will not only address renovation but also improvement of the energy performance of the existing buildings. Towards this end, energy retrofitting actions complemented by the use of information technologies to promote user behavioural change and local energy generation show a great potential to significantly benefit society. Those benefits are defined in terms of saving energy, money and emissions; increasing property value; creating jobs in the building sector and improving living conditions (and related economic savings for healthcare system).

In reference to building energy retrofitting, the technology is generally well established. In GrowSmarter, it has been observed that passive measures (upgrade of building thermal envelope) alone sometimes present very long paybacks in mild climates. The addition of active measures to passive energy retrofitting has evidenced to be a good alternative to shorten the payback time. In this sense, because the definition of the business model for building energy retrofitting interventions is a key factor to upscale the solution, several business models were tested in the project. Public–private agreements have shown a good acceptance by citizens in case of targeting multiowner private residential buildings since the public administration generates trust among neighbours in front of privately promoted campaigns.

The regulation in local generation and electricity self-consumption (and electricity markets in general) differs to a significant extent in European countries. Clear and harmonized regulation according to the current state of technology is required in order to fully deploy at commercial level distributed energy resources and the use of data from smart meters. In practice, it is still not possible to consider a single model throughout the European Union that allows the scalability of products and services associated to local energy generation and communities.

HEMS deployed in the project have shown high technical requirements (maintenance, data communication and technology obsolescence) but are considered a promising tool to engage building users towards an energy-efficient behaviour. However, in order to enhance the economic feasibility of these platforms, it is advisable to find synergies with other home services and look for opportunities to sell the platforms with integrated services packages.

Finally, the executed interventions within the GrowSmarter project have demonstrated that we need to put urban citizens at the centre of all the actions in order to succeed. Communication, information and user engagement is key to build trust, together with educating citizens and guaranteeing the correct implementation and operation of energy-efficiency measures. We need building users' awareness and acceptance to make low-energy districts the preferred development path for our society.

ACKNOWLEDGEMENTS

This research has been financially supported by the research and innovation programme Horizon 2020 of the European Union under the grant agreement nr. 646456 (GrowSmarter) as well as the GEIDI project (ref. TIN2016-78473-C3-3-R) financed by the Ministry of Economy and Competitiveness of Spain and the Generalitat de Catalunya (2017 SGR 1219).

REFERENCES

[1] United Nations, *World Urbanization Prospects: The 2014 Revision, Highlights,* Department of Economic and Social Affairs, Population Division, United Nations, 2014.

[2] European Commission. Smart cities, https://ec.europa.eu/info/eu-regional-and-urban-development/topics/cities-and-urban-development/city-initiatives/smart-cities_en (accessed 1 April 2019).

[3] European Commission. Horizon 2020, https://ec.europa.eu/programmes/horizon2020/en (accessed 1 April 2019).

[4] European Commission. COSME. Europe's programme for small and medium-sized enterprises. Internal Market, Industry, Entrepreneurship and SMEs. https://ec.europa.eu/growth/smes/cosme_en (accessed 2 April 2019).

[5] GrowSmarter, http://www.grow-smarter.eu/home/.

[6] European Commission. Smart Cities & Communities. Innovation and Networks Executive Agency, https://ec.europa.eu/inea/en/horizon-2020/smart-cities-communities (accessed 2 April 2019).

RESEARCH ON DESIGN METHOD FOR THE BLUE-GREEN ECOLOGICAL NETWORK SYSTEM TO DEAL WITH URBAN FLOODING: A CASE STUDY OF CHARLESTON PENINSULA

ZHITONG LIANG[1], ROBERT REID HEWITT[2] & YAN DU[1]
[1] Key Laboratory of Urban Agriculture in Central China, Ministry of Agriculture and Rural Affairs, Huazhong Agricultural University, China.
[2] Clemson University, United States of America.

ABSTRACT

The landscape strategy to deal with climate change has become an important issue in the process of sustainable urban development in the world. Particular focus is given to the Charleston Peninsula in South Carolina, USA, which faces floods due to inefficiency in stormwater collection systems, increased frequency of intense rain events, excessive impervious surfaces, tide cycles, etc. In addition, hurricane events and sea-level rise are considered sources of flood risk in the coastal areas of the peninsula. This research draws on existing urban stormwater management theory to argue that the blue and green water ecological network system in the built-up area represents an innovative approach to alleviate flooding and promote a healthy landscape during urban renewal. According to the analysis of hydrological characteristics, the peninsula is divided into 17 basins, and then each basin is studied separately. Within Basin 8, the potential block is divided into four types of functional stormwater management units (fast flow zone, absorption containment zone, additional digestion zone, and upstream interception zone) and connected by a reintegrated drainage system. Finally, the corresponding micro-landscape strategy is proposed according to the block property. Functional units simultaneously undertake the functions of rainwater management and landscape activities. In the end, the new active-recreation space and passive-recreation space in the network are connected with the original urban green space and provide the city with a series of unique ecosystem services to support urban drainage systems and human health. It is hoped this research will provide an attempt for future urban stormwater management from the perspective of landscape planning and design.

Keywords: blue-green ecological network system, hydrological process, landscape architecture, public space system, stormwater management unit.

1 INTRODUCTION

Although there are already mature examples of rainwater management, such as Best Management Practices (BMP), Sustainable Urban Drainage System (SUDS), and Water Sensitive Urban Design (WSUD), the research on the sustainable management of rainwater systems for climate change is still rising. Its essence is to control surface runoff, reduce water pollution, and restore the integrity of ecosystems [1]. Among them, stormwater control measures (SCMs) capture and retain rainwater through on-site infiltration, remove rainwater pollutants, and minimize rainwater runoff [2]. When SCMs are integrated into the basin in an appropriate manner, they do not affect the water quality of natural water bodies [3]. In addition, research on green infrastructure (GI) has shown that it has a positive effect on reducing stormwater runoff. Through different combinations of low-impact development strategies, rainwater runoff can be effectively reduced for individual land use and even for the entire basin [4].

Similar to many cities, Charleston's natural hydrological process was disrupted by urban development. However, its geographical location and special historical and social factors make it face more frequent and complicated flood disasters. Based on the characteristics of hydrological processes, this study attempts to build a stormwater management plan based on the blue and green ecological network system of the peninsula.

© 2020 WIT Press, www.witpress.com
DOI: 10.2495/DNE-V14-N4-275–286

2 BACKGROUND

The Charleston Peninsula is located at the entrance to the Atlantic Ocean in southern South Carolina, USA, and is surrounded by the Ashley River and the Cooper River (Fig. 1). Charleston has ample rainfall throughout the year, with approximately half of the annual rainfall concentrated between June and September. The rainy season has an average of 25 days of rainfall per month. In addition, under the influence of global climate change, extreme weather, such as hurricanes, is appearing at a higher frequency. Between 2015–2017, strong hurricanes entered the Charleston Peninsula three times, bringing heavy rainfall and storm surges up to 3 meters [5]. On sunny days, the tides only cause floods in a small area near the coast. However, as the sea level rises, the land area that is submerged by seawater during high tide expands further [6]. All of these factors have brought surface runoff to Charleston, which is the cause of floods. According to data from the National Oceanic and Atmospheric Administration (NOAA) of the United States, Charleston's nuisance floods have increased by 400% since the 1960s.

Surface runoff is accompanied by evaporation and infiltration that then flows into the drainage system or flows along natural terrain. The peninsula has a relatively flat terrain and complex microtopography that causes surface floods to remain and impedes effective surface water collection systems. Urban expansion has brought a large amount of impervious ground and laterally increased the peak surface runoff. At the same time and due to the long history of construction, the pipe sizes are not appropriate, and the drainage efficiency is greatly affected by the tide situation. This is far from meeting the needs of contemporary drainage.

As one or more of the factors listed in the current text for city surface runoff are superimposed, Charleston has to face frequent floods due to poor surface water supplies and inefficient drainage systems. In recent years, Charleston has placed great emphasis on flood management, focusing on the construction of engineering systems and GIs for low-impact development. However, the low-impact development measures focus on encouraging self-individual behavior: encouraging residents to increase the green area, achieving preliminary rainwater harvesting, and utilizing green roofs and rain gardens in private housing. There is a lack of systematic and effective planning at the city level. Based on this, the concept of the blue-green ecological network was proposed to explore the construction method of ecological rainwater-guiding landscape systems applicable to the Charleston Peninsula. Based on the characteristic analysis of the hydrological cycle process, flood management is carried out by means of landscape intervention to alleviate the urban flood pressure. At the same time, it complements the urban open-space system.

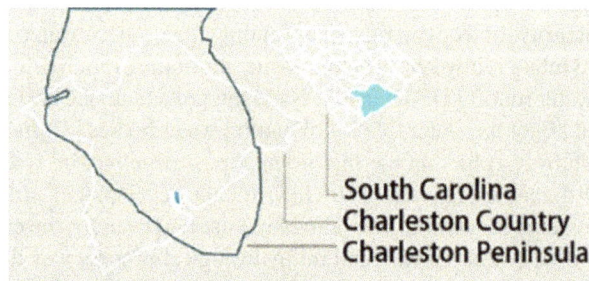

South Carolina
Charleston Country
Charleston Peninsula

Figure 1: Location of the Charleston Peninsula.

3 CHARLESTON PENINSULA CASE STUDY: SOURCES AND METHODS

The concept of the blue-green network system is not new and has been incorporated into general urban planning based on rain and flood management [7]. At the national level and even on a transnational scale, the ecological value of blue-green networks is very prominent, making it a kind of background ecological network [8–9]. This is a blue-green network concept from the perspective of landscape ecology. However, the blue-green network system in this study refers to an urban space network system formed by a combination of blue and green infrastructures. Urban stormwater management should emphasize small-scale restoration or conservation of natural hydrological processes [10]. When the concept of blue-green networks is applied to the urban watershed management target, how should this network be built? What is the basis of it? The construction of the network system in this study consists of three main steps: analysis of peninsula hydrological characteristics and basin division, reorganization of surface runoff confluence and establishment of stormwater management units for specific basin, and realization of landscape ecological functions (the second and third steps are summarized in section 4.2).

3.1 Peninsula hydrologic analysis and basin division

Data from government agencies' websites and documents were collected first, and the risk maps of different types of flood disasters (such as storm, tide, and storm surge) were sorted. Among them, floods caused by heavy rainfall are the most widespread and frequent disaster type. The construction of blue-green networks requires interdisciplinary research methods [11]. In the next step, GIS hydrological analysis was performed using DEM data with a precision of 3.0 M downloaded from the NOAA website to establish an overall understanding of the basic hydrological characteristics of the peninsula. Digital elevation models (DEM) are widely used in various environmental studies and are commonly used to obtain information on flow directions, flow accumulation, drainage networks, and watershed partitioning [12]. Hydrologic analysis tools in ARCGIS software are used to carry out filling analysis, define flow direction, capture river network outlets, extract flow network and flow network classification, etc.; a surface runoff classification diagram, runoff flow direction diagram, and basin zoning map automatically divided by the program can be obtained for the whole peninsula. The hydrological analysis results can solve two problems: (1) the flow direction and grade of surface runoff during rainfall, and (2) the division of the basin.

3.2 Construction of blue-green network for a specific basin

The result of basin division is the basis of further hydrologic research [13]–[14]. Artificial integration is also required for basins obtained in the previous step, which are automatically divided according to the outlet of the river network. This defines an area where tributaries eventually converge at the same outlet as a large basin. According to this principle, the small watershed is integrated, and finally the basin zoning result required for planning and design is obtained. Therefore, the confluence process of each basin is independent, which is the scientific support for the logic of subdivision management of peninsula flooding. One of the basins will be selected to demonstrate the construction of a blue-green network system in this section.

3.2.1 Basic information

Data on land cover types and flood distribution during the once-in-a-decade rainstorm events in the basin (hereafter referred to as the flood area) were mapped on the satellite map. Flood area and runoff maps generated by high-precision DEM were mutually verified to evaluate their reliability.

3.2.2 Analysis of comprehensive confluence process and flood detention reasons in the basin

The influence and function of underground drainage networks in urban basins cannot be ignored. At this stage, the analysis results of the SWMM model were used as an important reference for planning. SWMM developed in 1971 by the United States Environmental Protection Agency (USEPA, 2000), and is widely used in simulation, analysis, and design [15]. The National Oceanic and Atmospheric Administration (NOAA) has published an analysis of how drainage networks operate under various conditions using the SWMM model for similar regions of designated basins. In this paper, the analysis results are statistically collated and visualized on satellite images. It makes sense to have a clear understanding of the breakdown thresholds for the drainage system. This factor can be geographically calibrated and superimposed on the distribution of flood areas. The cause of the water accumulation can be preliminarily speculated based on these data; it may be due to the inability of runoff to reach the inlet or the fact that the runoff is far beyond the drain's load.

3.2.3 Rainwater management unit and surface confluence route

Flood areas, areas through which primary runoff flows (including adjacent areas), are divided into plots bounded by urban roads. After recording the statistics of land type, ownership, area, and other factors, the stormwater management potential of various plots was assessed subjectively. Stormwater management potential, location relationship between plots and runoff, and causes of flooding become the basis for defining rainwater management units (Fig. 2), which will be described in detail below. Urban landscape-based stormwater infiltration systems are typically modular, rely on a minimum of hard engineering, operate by gravity, and therefore are defined by local and microtopography [16]. The original runoff route was adjusted according to the distribution of various stormwater management units. The guidance of surface runoff mainly depends on microtopography and low-level engineering methods. The corresponding rainwater control function is given to the different land parcels, making it a stormwater management unit, and the units are connected to the restored water system to form a natural drainage network with both flood storage and flood discharge capacity.

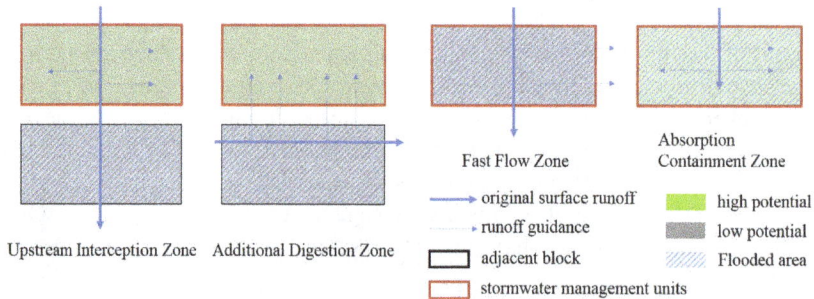

Figure 2: Four types of stormwater management units.

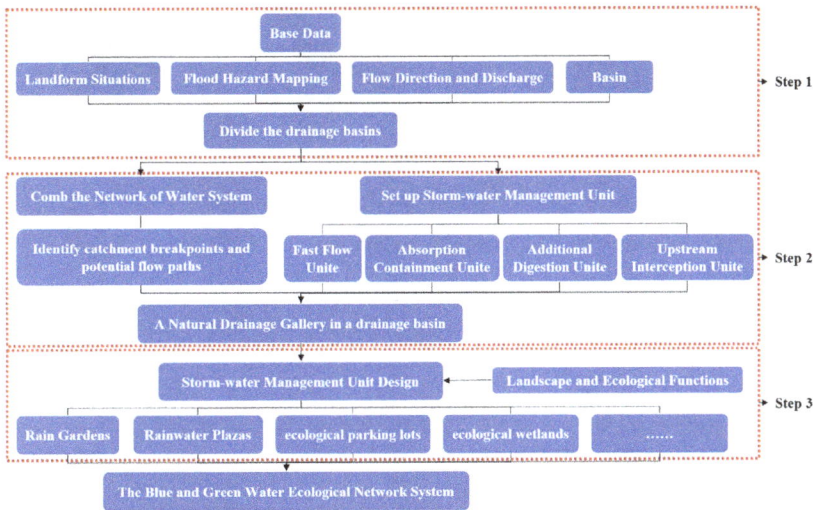

Figure 3: Blue-green ecological network construction flow chart.

3.3 Landscape intervention

Blue and green infrastructure often carries a natural scientific or technical interpretation which involves some limitations in perspective [17]. Research shows that focusing only on natural science and technological means creates unnecessary disputes among stakeholders [18]. The social and cultural value of blue-green networks should not be ignored. The intervention of landscaping means in the specific design process of rainwater management can effectively integrate the ecological value and humanistic value of a blue-green network. The blue-green ecological network construction process is shown in Fig. 3.

4 RESULT

4.1 Analysis

The effects of storm surges, tides, and rising sea levels on the entire peninsula are not evenly distributed in terms of space and intensity. Partial areas may also be threatened by several types of disasters [19]. The way to deal with floods should be more localized. Despite this, the peninsula is high in the middle and low on the edge, and it is obvious that the lower elevation coastal areas are more seriously affected.

On the GIS platform, the surface DEM data of the peninsula with an accuracy of 3.0M was used for the analysis of surface runoff extraction, flow direction analysis, and basin analysis. From the analysis results, it can be observed that several independent surface runoffs formed on the surface of the peninsula after rainfall, are flowing from the center to the coast (Fig. 4a and b). According to the characteristics of the catchment, 17 drainage basins were integrated from the preliminary basin zoning results obtained from the analysis (Fig. 4c). The surface coverage of each basin is very different [20]. Subsequent research is carried out according to the basin division. The natural drainage network of each basin not only undertakes the function of flood management but also is given the landscape and ecological functions and, finally, organically forms a blue-green ecological network of the entire peninsula.

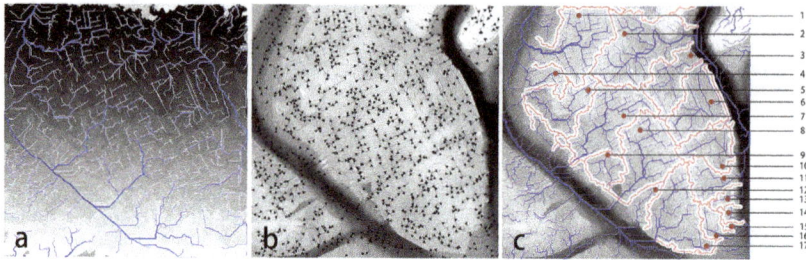

Figure 4: Hydrological analysis map of the peninsula. (a): Runoff extraction and classification; (b): Runoff direction analysis map; (c): Basin map of the peninsula.

4.2 Planning

The following section will explain the planning process from the construction of the blue network and the green network, so that the concept is better understood. The blue network includes guidance on the natural surface drainage process, and adopts specific measures, such as drainage, storage, and diversion for specific conditions. The green network means that the above water treatment is realized by landscape means to turn the threat into an opportunity. In order to better study the network, it is necessary to come down to a local scale of analysis. Taking Basin 8 as an example, the construction process of the blue-green ecological network of the basin and the micro-design of the stormwater management unit will be described in detail.

Basin 8 is located in the southwest of the peninsula. It is mainly composed of medium- and low-intensity development land but lacks large-scale green land. The largest body of water in Basin 8 is a triangular lake that is connected to the sea (Fig. 5a). After superimposing the runoff map with the 10-year flood basin map, it is found that the flooding area is mainly located at midstream and downstream of the runoff. It means the flood water volume has exceeded the load of both the pipeline drainage and natural drainage (Fig. 5b). However, there are still small flooding areas within Basin 8 where the main runoff does not flow. The presumed reason is that the local micro-topography makes the rainwater gather and cannot be smoothly discharged into the inlet, meanwhile rainwater cannot infiltrate, resulting in water accumulation.

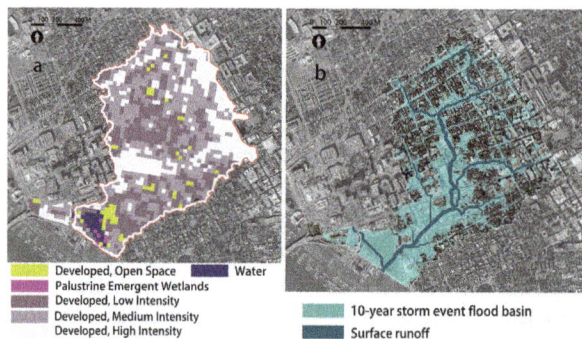

Figure 5: Basic information of Basin 8. (a): Land cover map of Basin 8. (*Compilation by the authors from the sources [20].*); (b): Runoff and the 10-year flood basin map of Basin 8. (*Compilation by the authors from the sources [21].*)

The Stormwater Management Model (SWMM) is a dynamic precipitation-runoff simulation model used to simulate a single or long-term precipitation event or water quality simulation in a city. The model is now well developed and widely used. NOAA uses this model to evaluate the performance of existing rainwater harvesting systems in the Calhoun West service area of Charleston and simulate the combined effects of coastal flooding and precipitation to pinpoint when and where the stormwater system is at risk of being compromised (Calculation Example: Impacts of Coastal Flooding on Stormwater Infrastructure – City of Charleston, South Carolina).

The study also divides the study area into several basins, each of which sets a node, and assumes that all surface runoffs within the basin eventually flow into the underground pipeline through the node. The analysis results are summarized as follows (the simulation results of Scenario 3 are not related to this study and are not mentioned). On the basis of the existing sea level in Scenario 1, if there is no tidal backflow valve in the pipeline network, seawater overflow will occur when the tide is high. Based on the rising sea level 25 years later (2043) for Scenario 2, most areas of the pipeline cannot meet drainage needs in the event of a ten-year storm. Scenario 3 assessed the most extreme situation. The condition at this time is that the sea level has risen after 25 years, and the peninsula faces both a 25-year storm and a 2-year rainfall runoff event. The distribution of pipeline flooding nodes is the same as that in Scenario 2, but the flooding duration is longer and the water depth is greater (Fig. 6).

Considering the actual working conditions of the drainage channel of Basin 8 and the surface flooding situation at the same time is conducive to a better analysis of the causes of water accumulation, so that flood management can be more specific. Therefore, the results of the above SWMM model analysis in Scenario 2 were geo-calibrated with Basin 8. The area of the drainage overload in Basin 8 is then classified according to the degree of overload (Fig. 7a). There are many cases where the surface catchment watershed does not completely overlap with the underground pipeline service area (Fig. 7b). This does not affect the current analysis results.

Nodes such as J0.5, J1, J12, J41, and J42 predicted no overloading, but the 10-year flood map showed small floods in the service areas of these nodes. Therefore, the accumulation of water in these areas is likely due to poor water supply and the inability to smoothly discharge into underground pipelines. These areas are upstream of surface runoff, where the goal of flood management is to mitigate downstream drainage pressures without being flooded. The N10, N11, N12, and other nodes are overloaded with water for about 4 hours, but only some areas of the service area are flooded. One way to defer this situation is to divert the flood and

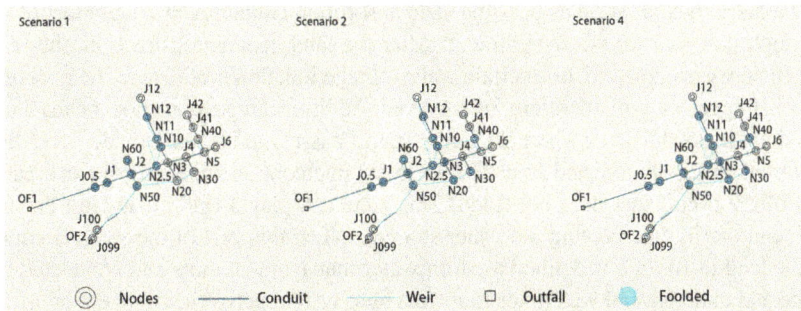

Figure 6: Flooding nodes in Scenario 1, Scenario 2, and Scenario 4. (*Compilation by the authors from the sources [22].*)

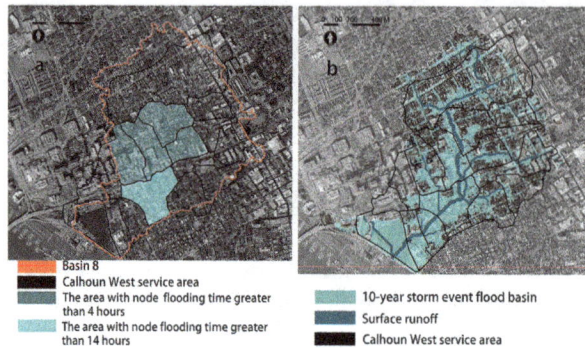

Figure 7: Comprehensive analysis map of hydrological processes. (a): The area of the drainage overload in Basin 8. (*Compilation by the authors from the sources [22].*); (b): Calhoun West service area geo-calibrated with runoff and 10-year event flood basin map. (*Compilation by the authors from the sources [21]–[22].*)

to avoid the peak hour. Node N50 is flooded for up to 14 hours, and its service area is located in a severe flooding zone just like the other downstream areas. However, these areas also have a certain scale of green spaces, wetlands, and water bodies. The main strategy for flood management of these areas should be to accelerate emissions while strengthening the temporary water storage function in partial areas. Combined with the flood management strategies for reducing the instantaneous flow of floods in other upstream areas, the flood damage intensity can be reduced and the flooding time shortened.

The first step in the planning section is to identify blocks with potential in Basin 8 based on two factors: surface runoff and 10 years of rainstorm flooding. These blocks are then classified according to residential areas, hard squares, roads, parking lots, green spaces, etc. (Fig. 8a). In particular, it is important to note that the level of potential depends on the condition of the underlying surface and the size of the area that can be modified. The second step is to divide the land into different stormwater management units based on the discussion of the three main governance strategies for different regions mentioned above. At the same time, the surface runoff is reorganized based on the original flow. The main guiding methods are microtopography reconstruction and the addition of low-impact means, such as culverts (Fig. 8b). There are four types of stormwater management units: fast flow zone, absorption containment zone, additional digestion zone, and upstream interception zone. Fast flow zone refers to the area where rainwater needs to be discharged as soon as possible. The absorption containment zone refers to the plot having a high potential for water treatment. After the landscape reconstruction, the rainwater seepage efficiency is improved, or a certain water storage function is realized. The goal of reducing runoff in this region will, therefore, be achieved. Additional digestion zone means the land is not located in the floodplain or where the major runoff passes, and it can provide flood diversion services for other heavily affected area. The upstream interception zone refers to an area located upstream of the runoff that does not flood. This zone can play a role in slowing downstream peaks by temporarily intercepting the water storage. When this part of the work is completed, most of the land in Basin 8 is defined as stormwater management units and connected through the surface and underground waterlines to form a blue system network within the area.

Within each stormwater management unit, corresponding landscape measures are proposed based on the current status of the land use, including rain gardens, rainwater plazas,

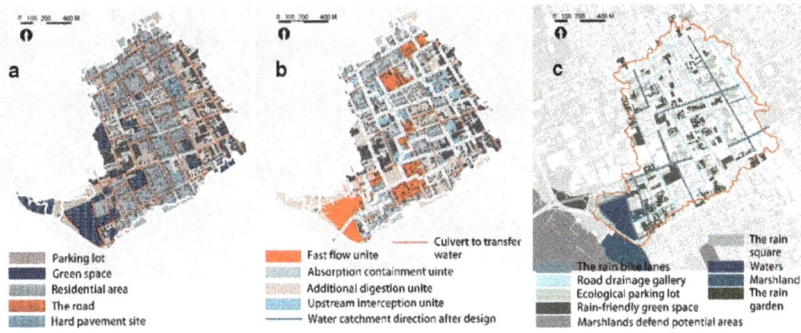

Figure 8: Map of the planning process. (a): Potential block classification map; (b): Stormwater management unit and reorganized surface runoff; (c): Landscape measures map of stormwater management.

ecological parking lots, rain-friendly green spaces, ecological wetlands, road drainage corridors, and rain bicycle lanes where the upper plan of the bicycle lane is used as the basis for road selection (Fig. 8c). Most of the land types in Basin 8 belong to residential areas that have a certain area of hard paved ground or green space. The rain garden is small and flexible and can be easily placed in the gaps between housing constructions within these areas. In addition, the total number of parking lots and hard paved plazas is large with sufficient space to build underground reservoirs and other water management facilities. Meanwhile, the parking lot was integrated into this opportunity to improve land use efficiency. Under the premise that the total number of parking spaces is not reduced, more land is replaced by community open space. Terrain adjustments, vegetation modifications, and other strategies are applied to natural green spaces, such as parks and wetlands, to improve rainwater regulation and purification. Water-compatible new public spaces, such as rainwater theme parks and floating sport fields, have also been proposed; the treated rainwater can be used as landscape water in these places. Roads are an important element in connecting the rainwater control unit. The control methods adopted for some roads with poor drainage include: replacement of permeable pavement, micro-slope matching grass drainage ditch, etc. The runoff on the road eventually enters the underground drainage system or the underground purification and storage device installed in the green space and parking lot. The planned bicycle lanes should combine function of use with rainwater management function. In this way, a green network has been built.

4.3 Map details

Near the southwest side of the site, two green spaces outside Basin 8 are presented on maps. The one on the north is the green space under the highway, and the other one is the wetland connected to the Ashley River. According to field research, the green space under the highway is in a state of ruin. It is located downstream of the surface runoff of Basin 9 and is close to the flood-hit area of Basin 8. Therefore, it has important regulation and storage impacts for both Basin 9 and Basin 8. According to the runoff analysis of the entire peninsula, the runoff of Basin 9 eventually enters the triangular lake of Basin 8 (Fig. 4c). It is recommended that the runoff of Basin 9 be diverted, so that it eventually passes through the green space under the highway to alleviate flooding in Basin 8. This space can also serve as a temporary storage block for Basin 8.

Basin 8 is also threatened by storm surges. During these disasters, there are huge waves from the sea in addition to the heavy rainfall. According to records, the maximum wave height caused by a hurricane in Charleston between 2015 and 2017 was about 9 feet [5]. The infrastructure, houses, and even the lives of residents near the coast are threatened by huge waves. Therefore, the wetland between the Ashley River and the peninsula has an important buffering effect. Using organic means, such as capturing and accumulating silt and cultivating oyster reefs, is planned to promote the organic growth of narrow wetlands and form an effective wave buffer zone.

5 DISCUSSION

People's activity spaces can be divided into active-recreation spaces and passive-recreation spaces. Active-recreation space refers to the destinations, such as city parks, where the residents have clear travel needs and have time constraints. Passive-recreation space refers to the space in the process of the residents' work, daily life, commuting, and so on. Most of the southern end of the Charleston Peninsula is a historical block. The land type is relatively simple and lacks sufficient open space. Through this plan, the public space system of the peninsula can be established. Taking Basin 8 as an example, the reconstructed green spaces, wetlands, and other spaces can form a new type of high-quality active-recreation space. Spaces like community rain gardens, ecological parking lots, and rain bike lanes act as passive-recreation spaces, connecting scattered open spaces to form a complete network. Therefore, the blue-green ecological network proposed in this study is a functional composite spatial configuration for the Charleston Peninsula. It is also a strategic means to reconcile multifaceted issues such as disaster, ecology, and space utilization.

This research mainly involves the process of network construction and provides an idea for overall planning and regulation. The previous hydrological analysis stage has certain limitations in terms of data and analysis methods, and the performance problems after completion have not been considered. Blue and green infrastructure often carries a natural scientific or technical interpretation, which involves some limitations in perspective [23]. Collaborative planning is needed to integrate urban retrofitting, development processes, and flood risk management [24]. A robust framework requires a combination of four vulnerability reduction functions: threshold capacity, coping capacity, recovery capacity, and adaptive capacity [25]. This study takes the Charleston peninsula as an example to describe one of the physical space construction processes of the blue-green network system. In practical application, factors such as evacuation shelters, social capital, repair, and reconstruction should also be considered.

The extraction of runoff direction and the division of the watershed are based on the basic analysis method of hydrology. Urban hydrology is easily affected by land use and infrastructure, and it is still a complex task to determine drainage networks and watershed boundaries in highly urbanized river basins [26]. Increasing numbers of tools, such as Geo-PUMMA, are being developed to simulate urban or suburban hydrological processes. Similar research will also become more practical with the further optimization of analytical tools. It is reported that government agencies and academic institutions are very concerned about the floods in Charleston and are conducting more in-depth research. It is believed that future research results of the above departments can provide useful scientific support for the subsequent urban planning and design.

ACKNOWLEDGEMENTS

We should like to thank Professor Robert R. Hewitt and Hala F. Nassar (Clemson University) for their help in Collecting basic information and Field investigation. We also want to thank

Professor Li Zhang and Wei Zhang (Huazhong Agricultural University), for their valuable comments. This research was funded by the Creative Research Program of DDON Planning&Design Inc.

REFERENCES

[1] Shishegar, S., Duchesne, S. & Pelletier, G., Optimization methods applied to stormwater management problems: A review. *Urban Water Journal*, **15(3)**, pp. 1–11, 2018. https://doi.org/10.1080/1573062x.2018.1439976

[2] Sadeghi, K.M., Loáiciga, H.A. & Kharaghani, S., Stormwater control measures for runoff and water quality management in urban landscapes. *JAWRA Journal of the American Water Resources Association*, **54(1)**, pp. 124–133, 2018. https://doi.org/10.1111/1752-1688.12547

[3] Rivers, E., McMillan, S., Bell, C. & Clinton, S., Effects of urban stormwater control measures on denitrification in receiving streams. *Water*, **10(11)**, p. 1582, 2018. https://doi.org/10.3390/w10111582

[4] Eaton, T.T., Approach and case-study of green infrastructure screening analysis for urban stormwater control. *Journal of Environmental Management*, **209**, pp. 495–504, 2018. https://doi.org/10.1016/j.jenvman.2017.12.068

[5] Historic Events; Charleston, SC—Official, https://www.charleston-sc.gov/2007/Historic-Events

[6] Trees to Offset Stormwater, Case Study 04: Charleston, South Carolina; SC Forestry Commission, Green Infrastructure Center, the City of Charleston, Forest Service Department of Agriculture, https://www.charleston-sc.gov/DocumentCenter/View/19091/Trees-and-Stormwater-Study-Charleston-SC-2018?bidId=, (accessed August 2018)

[7] Wagner, I., Krauze, K. & Zalewski, M., Blue aspects of green infrastructure. S*ustainable Development Applications*, **4**, pp. 145–155, 2013. https://doi.org/10.1016/j.jenvman.2017.12.068

[8] Hong, S.K., Nakagoshi, N. Fu, B.J. & Morimoto (eds.), Y., Landscape ecological applications in man-influenced areas. *Landscape Ecology*, **23(10)**, pp. 1291–1292, 2008. https://doi.org/10.1007/s10980-008-9286-8

[9] Hugo, P., The network approach: Dutch spatial planning between substratum and infrastructure networks. *European Planning Studies*, **15(5)**, pp. 667–686, 2007. https://doi.org/10.1080/09654310701213962

[10] Burns, M.J., Fletcher, T.D., Walsh, C.J., Ladson, A.R. & Hatt, B.E., Hydrologic shortcomings of conventional urban stormwater management and opportunities for reform. *Landscape and Urban Planning*, **105(3)**, pp. 230–240, 2012. https://doi.org/10.1016/j.landurbplan.2011.12.012

[11] Lawson, E., Thorne, C., Ahilan, S., Allen, D., Arthur, S., Everett, G., ... & Kilsby, C., Delivering and evaluating the multiple flood risk benefits in blue-green cities: An interdisciplinary approach. In *Flood Recovery Innovation and Response IV*, eds. D. Proverbs & C.A. Brebbia, **184**, pp. 113–124, 2014. https://doi.org/10.2495/friar140101

[12] De Sousa, T.M.I. & Da Paz, A.R., How to evaluate the quality of coarse-resolution DEM-derived drainage networks. *Hydrological Processes*, **31(19)**, pp. 3379–3395, 2017. https://doi.org/10.1002/hyp.11262

[13] Bell, C.D., Mcmillan, S.K., Clinton, S.M. & Jefferson, A.J., Hydrologic response to stormwater control measures in urban watersheds. *Journal of Hydrology*, **541**, pp. 1488–1500, 2016. https://doi.org/10.1016/j.jhydrol.2016.08.049

[14] Jankowfsky, S., Branger, F., Braud, I., Gironás, J. & Rodriguez, F., Comparison of catchment and network delineation approaches in complex suburban environments: application to the Chaudanne catchment, France. *Hydrological Processes*, **27(25)**, pp. 3747–3761, 2013. https://doi.org/10.1002/hyp.9506

[15] Kong, F., Ban, Y., Yin, H., James, P. & Dronova, I., Modeling stormwater management at the city district level in response to changes in land use and low impact development. *Environmental Modelling & Software*, **95**, pp. 132–142, 2017. https://doi.org/10.1016/j.envsoft.2017.06.021

[16] Ahern, J., Urban landscape sustainability and resilience: The promise and challenges of integrating ecology with urban planning and design. *Landscape Ecology*, **28(6)**, pp. 1203–1212, 2012. https://doi.org/10.1007/s10980-012-9799-z

[17] Assmuth, T., Hellgren, D., Kopperoinen, L., Paloniemi, R. & Peltonen, L., Fair blue urbanism: demands, obstacles, opportunities and knowledge needs for just recreation beside Helsinki Metropolitan area waters. *Urban Sustain*, **9(3)**, pp. 253–273, 2017. https://doi.org/10.1080/19463138.2017.1370423

[18] Kati, V. & Jari, N., Bottom-up thinking—Identifying socio-cultural values of ecosystem services in local blue–green infrastructure planning in Helsinki, Finland. *Land Use Policy*, **50**, pp. 537–547, 2016. https://doi.org/10.1016/j.landusepol.2015.09.031

[19] NOAA Coastal Flood Exposure Mapper; NOAA Office for Coastal Management, https://coast.noaa.gov/digitalcoast/tools/flood-exposure, (accessed 26 April 2019)

[20] Data Access Viewer; NOAA Office for Coastal Management, Online, https://coast.noaa.gov/dataviewer/#/lidar/search/

[21] Calhoun West Drainage Improvement & Sea Level Rise Mitigation Project, Watershed Assessment; City of Charleston, https://www.arcgis.com/apps/MapJournal/index.html?appid=7cd50ff336e04e44820bec01f816a9d5

[22] Analyze Stormwater Systems, Calculation Example: Impacts of Coastal Flooding on Stormwater Infrastructure—City of Charleston, South Carolina; NOAA Office for Coastal Management, https://coast.noaa.gov/stormwater-floods/analyze/

[23] Assmuth, T., Hellgren, D., Kopperoinen, L., Paloniemi, R. & Peltonen, L., Fair blue urbanism: Demands, obstacles, opportunities and knowledge needs for just recreation beside Helsinki Metropolitan area waters. *Urban Sustain*, **9(3)**, pp. 253–273, 2017. https://doi.org/10.1080/19463138.2017.1370423

[24] Van Herk, S., Zevenbergen, C., Ashley, R. & Rijke, J., Learning and action Alliances for the integration of flood risk management into urban planning: A new framework from empirical evidence from the Netherlands. *Environmental Science and Policy*, **14(5)**, pp. 543–554, 2011. https://doi.org/10.1016/j.envsci.2011.04.006

[25] De Graaf, R.E., van de Ven, F.H.M. & van de Giesen, N.C., The closed city as a strategy to reduce vulnerability of urban areas for climate change. *Water Science and Technology*, **56(4)**, pp. 165–173, 2007. https://doi.org/10.2166/wst.2007.548

[26] Kayembe, A. & Mitchell, C.P.J., Determination of subcatchment and watershed boundaries in a complex and highly urbanized landscape. *Hydrological Processes*, **32(18)**, pp. 2845–2855, 2018. https://doi.org/10.1002/hyp.13229

EFFECTS OF RELATIVE EFFICIENCY AND INDUSTRIAL DIVERSITY ON PRODUCTION OF OLD INDUSTRIAL COMPLEX

MYOUNG SUB CHOI[1] & HWAN YONG PARK[2]
[1]Government Innovation & Productivity Institute, The Catholic University of Korea, South Korea
[2]Urban Planning, Gachon University of Korea, South Korea

ABSTRACT
This study aimed to evaluate the effects of relative efficiency and industrial diversity of old industry complex on production. Cobb–Douglas production function was estimated with consideration of relative efficiency and inverse of Herfindahl–Hirschman Index for the 94 old industrial complexes during 2014–2017. The effects on production which would be varied by industrial complex types and location types were also considered in the production model. As a result, statistically significant positive effects on productivity in old industrial complex have been estimated regardless of not only types of industrial complex (national and general industrial complex) but also location type (capital and non-capital area). In contrast, diversity estimated has a negative impact on productivity, but it does not show statistical significance. Therefore, to activate old industrial complex, plans for increases of relative efficiency by operation cost reductions of businesses in industrial complex will be needed. And to diversify the industrial types in old industrial complex, plans should consider the types of industrial complex and location type. Industrial linkages among companies in old industrial complex should also be considered in the process of selecting business.
Keywords: relative efficiency, industry diversity, Herfindahl–Hirschman index, Cop–Douglass production function, deteriorated industrial complex

1 INTRODUCTION
The revitalization of old industrial complexes is important because they affect not only the growth of the industrial complex but also the regional economic development. In addition, there are national benefits in that the activation of an existing industrial complex can minimize social costs, compared to the development of a new industrial complex. The government has responded to the problem of aging industrial complexes and is trying to improve the competitiveness of aging industrial complexes by rebuilding industrial complexes, improving the structure and enacting special laws. On the other hand, there seems to be a lack of examination of the production side of the old industrial complexes or the industrial complexes itself. Regarding the existing industrial complexes, they still analyse the determinants of competitiveness, location, price and employment. It was not enough. Therefore, this study focuses on the relative efficiency and industry diversity among these determinants.

The relative efficiency of old industrial parks is an indicator of how efficiently they operate, which can determine whether efficient industrial parks can lead to increased production. The diversity of industries in old industrial parks puts a question on whether the concentration of industries within an industrial park on specific industries helps production or whether the composition of various industries has a positive effect on production.

Hence, this study examines how the relative efficiency of old industrial parks and the diversity of industries affect their production. This study is carried out as follows. Chapter 2 examines how relative efficiency and diversity relate to production and presents differences

© 2020 WIT Press, www.witpress.com
DOI: 10.2495/EI-V3-N2-120-131

of the study. Chapter 3 describes the Cop–Douglas function used in the study, the relative efficiency and the analysis data to measure industry diversity. Chapter 4 illustrates the impact of relative efficiency and industry diversity on production, while Chapter 5 outlines the implications and limitations of research.

2 LITERATURE REVIEW

2.1 Relative efficiency and relationship with production

The relative efficiency measurement of industrial parks has been carried out in previous studies, mainly through data development analysis (DEA). Results from previous studies show relative efficiency differences in the size of the tenant companies, location characteristics, type of industrial complexes and operation period. And, to improve relative efficiency in the old industrial complex, they proposed to enhance operating rates, remodel aged facilities, increase external effects, and diversify industries [1–3]. However, previous studies have mainly focused on measuring the relative efficiency of industrial parks. In some studies, the determinant of relative efficiency was taken into account. Kim and Cho [2] measured relative efficiency through DEA for industrial park tenants and non-location enterprises, which were derived through the Tobit model.

Only a few studies take into account the approach to the effects of relative efficiency on production that are addressed in this study. Choi [4] analysed how the efficiency of the local government in the development region affected the growth of the local economy. The analysis showed that efficiency had a positive effect on economic growth. However, no approach has been made in terms of industrial parks. Therefore, this study will figure out the relationship between relative efficiency and production. As previously stated, relative efficiency is not necessarily a proportional relationship with production because it can be resolved in a way of reducing input costs in the same production scale [3].

2.2 The relationship between industrial diversity and production

The diversity of industries is related to the urban economy, and existing studies also explain the local economy, including its factors as an industrial structure or an urban economy. Byeon [5] analysed the effects of the integrated economy on the information and communication manufacturing industries in the Seoul metropolitan area and showed that the diversification of the industries had a negative effect on the aggregate. Mo and Kang [6] analysed the impact of regional industrial structures on the local economy. They found that industrial diversity has a negative effect on the national economy. But in terms of areas, the Seoul metropolitan area and Yeongnam region have a positive effect, and the Chungcheong and Jeolla provinces have a negative effect. In Park [7], for the entire manufacturing industry, the diversity of the industry was negative, but not statistically significant. However, in case of transport manufacturing, the effect was shown to be statistically negatively significant.

The existing studies show different effects of diversity and production depending on the analysis target, region and time. It comes on a background that there exists a limit to the assessment of how diversity in industries within industrial parks will affect production. For this reason, no research has been conducted only on industrial parks. In this aspect, this study is particularly different from existing research on the manufacturing industry field.

2.3 Difference in this study

Previous studies have led to a number of measures of the relative efficiency of industrial parks, but have not been conducted on whether relative efficiency helps increase production. Even though the concept of relative efficiency is reviewed, it is difficult to determine in which direction the effects of relative efficiency on production will be in an old industrial park, since the effects of higher relative efficiency in an old industrial park can be achieved by reduced input costs or increased production.

In comparison, there has been relatively much progress regarding the impact of industrial diversity on production, while focus has been on regional aspects rather than on industrial complexes. However, the effects of diversity on the local economy are not consistent with codes and statistical significance, which vary by analysis target, region and time. Therefore, it is interesting to check out how diversity in the industry will affect production in the aged industrial parks subject to this study.

Therefore, this study needs to take a concrete approach to how relative efficiency and industry diversity affect the production of old industrial complexes. For this purpose, the analytical model includes two variables discussed earlier in the production function approach, which was primarily used in existing studies.

3 FRAME OF ANALYSIS

3.1 Analysis method

The Cobb–Douglass production function is applied to analyse the effects of the efficiency of old industrial parks and the diversity of industries on production. The Cobb–Douglas production function is an economic model that can analyse the impact of production elements on production or value added, as it has previously been mainly used as one of the models for explaining production factors, especially in manufacturing [8–11].

The following production functions are set as the basic model. Here, Y, L, K and A represent industrial park output, number of workers, capital stock and total component productivity, respectively. However, in this study, the site area was used as a proxy variable for the capital stock, because it is not provided by the statistics related to the industrial park. Considering these limitations, Kim and Choi [11] also applied the land area as a proxy variable for the capital stock in estimating the production function of the old industrial park.

$$Y = AL^{\alpha} K^{\beta} \tag{1}$$

It is then assumed that total component productivity (A) consists of Dea (relative efficiency) and Div (industrial diversity) and other (B). Since the total element productivity reflects the remaining factors not included in single-element productivity measures, such as labour and capital, it is desirable to define relative efficiency and industry diversity as components of total element productivity in this study, which in turn affects production through changes in total element productivity.

$$A = BDe^{\gamma} Div^{\delta} \tag{2}$$

Combining the two formulas allows the final formula to be set as follows: the size of $\alpha+\beta$ means the economic level of scale and $\alpha+\beta>1$ means that the output will be greater than 1%, when the input of labour and capital increases by 1%, in which case the economy of scale will exist. Using the estimation results of the formula, it is possible to determine the level of economy of scale.

$$Y = BL^{\alpha}K^{\beta}Dea^{\gamma}Div^{\delta} \tag{3}$$

However, Formula 3 is a nonlinear model, which requires linearization of functions to be estimated by means of logarithmic transformation [12].

$$\ln Y = \ln B + \alpha \ln L + \beta \ln K + \gamma \ln Dea + \delta \ln Div \tag{4}$$

The relative efficiency index, one of the major independent variables, is derived from DEA. Previous studies have also estimated relative efficiency through DEA for industrial parks [1–3]. Efficiency in the DEA is defined as relative efficiency and is calculated between 0 and 1 because it represents the level of that Decision Making Unit (DMU) relative to the most efficient observation (DMU) [13]. In DEA, relative efficiency can be estimated separately on input and output basis from Charnes, Cooper and Rhodes (CCR) and Banker, Charnes and Cooper (BCC). This study is based on the BCC model on the output basis. The reason for applying the production approach in this paper is that developed industrial parks are easier to control production volume than inputs, and that BCC is more realistic than CCR that assume a constant return to scale [3].

The diversity of industries within old industrial parks uses the reciprocal of the Herfindahl–Hirschman index (HHI). The HHI has been used primarily to measure the diversity of the industrial structure at the urban and regional levels [14–16]. Therefore, it is also possible to apply it to the industrial parks. The HHI of old industrial parks ought to be the maximum value of 1 if all workers in the industrial park are concentrated in one industry, and the highest value will be $1/n$ if the number of workers is equally distributed in n industries. Therefore, the HHI is measured at a smaller value as the region's industrial structure becomes more diverse. The larger the HHI, the less diversity is interpreted. For the convenience, its reciprocal is used. Therefore, the larger the reciprocal value of the HHI, it can be interpreted as more diverse. This means that higher the diversity, the greater the output, if the estimated coefficients are positive.

$$\frac{1}{HHI_j} = 1 / \sum_{i=1}^{I}\left(\frac{E_{ij}}{E_j}\right)^2 \tag{5}$$

3.2 Analysis data

This study focuses on old industrial parks and targets 94 old industrial complexes derived from Park and Park [17]. The old industrial complex consists of 21 national industrial complexes and 73 general industrial complexes by type, while there are 30 metropolitan areas and 64 non-capital areas by location.

For the construction of models in this study, the production amount, site area, number of workers, relative efficiency index and the HHI are required. The basic data for estimating the production function model is based on the current status of the national industrial complex of the Korea Industrial Complex Management Corporation. The data provide the amount of production, land area and number of workers for each individual industrial complex. However, time span is extended to 2014–2017 to ensure statistical significance of the estimate. The production amount should be on a point-in-time basis in the production function model; therefore, the difference exists in value of the amount. The figure was adjusted to the base price in 2015 using the GDP deflator provided by the Bank of Korea's economic statistics system. The same approach is applied to DEA for the measurement of relative efficiency. The relative efficiency of this study is analysed as DMU for 94 individual aged industrial complexes, based on the BCC model which used site area and number of workers as input variables. Similar procedures for the analysis was used as in Choi [3].

Next, the HHI for measuring the diversity of industries is built using the raw data in individual industries of the Korea Industrial Complex Management Corporation between 2014 and 2017. The data provide key industries and the number of employees of individual companies in each industrial complex, so it can be applied to 94 aged industrial parks subject to this study. The HHI by industrial complex is based on the number of workers, not the number of companies. This is because the number of workers is more appropriate than the number of businesses to show diversity. Finally, the reciprocal of the derived HHI is included in the model. As of 2017, the correlation between the number of businesses, the number of workers and the amount of production in Korea's industrial parks was estimated, and the correlation between production and the number of employees (0.742) was larger than the correlation between production and number of businesses (0.570).

The basic statistics of the analysis data compiled are shown in the following table. The analysis data were divided into types of industrial parks (national and general industrial parks), location types (capital and non-capital areas) and two types of intersections (national and general industrial parks in capital areas, national and general industrial parks in non-capital areas). The results of production scale (production amount, area and number of workers) to determine the reliability of the construction data show that the national industrial complexes and non-capital areas are shown to be high. A similar pattern is also shown in the type of industrial complexes by location. Considering that national industrial parks are generally larger in terms of physical size than general industrial parks, it is believed that the data building in this study is reasonable in terms of the scale of production.

Next, the outcomes of relative efficiency are similar to those of size of production, in that the figures are high in national industrial parks and metropolitan areas. This is judged that relative efficiency of national industrial parks is likely be larger than that of general industrial parks because large enterprises or large companies with large production sizes are likely to move in. It is confirmed by the previous studies. Kim [18] showed the result that efficiency increases as a company grows in size and Noh [19] analysed that the relative efficiency of large companies is greater than that of small businesses.

However, the results of location type show the opposite results of production scale. This seems to be because older industrial parks located in capital areas have a higher effect of technological advancement compared to non-capital areas, as pointed out by Lee and Ahn [20]. The results of these elementary statistics are expected to have a positive impact on production by relative efficiency.

Finally, the results of industrial diversity are the opposite to the results of production scale, in general industrial parks and the capital areas, showing great diversity. A similar pattern was found by location type. The fact that national industrial parks are less diverse in industry than general industrial parks can be interpreted that national industrial parks are aggregated by similar industries on a division basis than general industrial parks. However, when subdivided into different types of industrial complexes by location type, the diversity of national industrial parks is high, and general industrial parks are high in capital areas. In view of this, it can be presumed that the impact of industrial diversity on production in the analysis of future production functions is basically negative, but the difference between location type and industrial complex type is likely to occur.

4 ANALYSIS RESULTS

4.1 Overall analysis

Estimation has been made, based on production functions, to review how relative efficiency and diversity affect the production of old industrial parks. The overall analysis of 94 aged industrial parks covered by this study is shown in Table 2. It was analysed that the sum of the regression coefficients of area and number of workers was 1.066, achieving economy of scale. In other words, aged industrial parks would see a 1.066% increase in production if the production element (area and number of workers) increased by 1%, showing a greater proportion of the increase in production to the increase in inputs, which is more than 1. Specifically, labour (the number of employees) has a greater impact than capital (the area). This is consistent with the results of existing studies on aged industrial park in certain areas, in which labour elasticity is greater than capital elasticity [11,21].

Next, the results of relative efficiency and industry diversity, which are the main points of this study, are as follows: relative efficiency was analysed to have a statistically significant positive effect on the output. This means that relative efficiency affects not only the efficient operation of old industrial parks, but also the economic growth of old industrial parks. However, it is found that the diversity of the industry (the inverse of the HHI) has a statistically negative effect on the production of old industrial complexes. This can be interpreted as a decrease in the total output of old industrial complexes as industries in old industrial complexes are diversified. This is viewed to be due to the large diversity, in general industrial parks and capital areas, with relatively small output, as indicated in the basic statistics summarized in Table 1.

Finally, the results according to location type and industrial complex type are as follows. The regression coefficient of dummy variable for the capital areas was negative (−). This appears to be due to the larger production volume of non-capital areas rather than capital areas, given the 94 aged industrial parks under analysis (see Table 2). However, what is unique is that the code of general industrial parks was presumed to be (+). This means that the basic output of general industrial parks is greater than that of national industrial parks, when controlling the production factors (area and capital), relative efficiency and diversity of industries, which was not derived from the results of the basic statistics in Table 1. These results suggest that

Table 1: Basic statistics.

		Production amount (KRW 100 million)	Area (thousand m²)	Workers (thousands)	Relative efficiency	Herfindahl reciprocal number
National		268,224	27.4	52.8	0.481	0.172
General		22,108	1.3	5.5	0.337	0.238
Capital areas		53,524	5.5	22.1	0.432	0.240
Non-capital areas		88,139	7.9	13.3	0.340	0.215
Capital areas	National	222,262	25.7	94.3	0.556	0.145
	General	11,339	0.4	4.0	0.401	0.265
Non-capital areas	National	286,609	28.0	36.2	0.451	0.182
	General	27,383	1.7	6.2	0.306	0.226

Table 2: Full analysis results.

	Coefficient	Standard error	t value	p value
(constant)	6.776	0.180	37.71	0.0001
ln area	0.427	0.025	17.05	0.0001
ln number of workers	0.639	0.022	28.53	0.0001
ln DEA	0.816	0.027	30.76	0.0001
ln Herfindahl reciprocal	−0.071	0.040	−1.76	0.0790
Dummy for metropolitan	−0.0178	0.052	−3.42	0.0007
Dummy for general parks	0.282	0.067	4.19	0.0001
Modified R^2	0.967			

Note: The reference group for general industrial park is the national industrial park and the reference group for metropolitan area is the non-capital area.

the production function of aged industrial parks should be estimated separately by the type of industrial park. This suggestion comes from one of the basic assumptions in estimating the overall production function that each of the independent variables has the same effect on production for each type of industrial complex. In particular, if the size of the capital elasticity and labour elasticity embodied in the production component is reversed by the type of industrial complex (capital elasticity), the analysis may be difficult to interpret, based on the assumption that its influence is the same for each type of industrial complex (such as labour elasticity and labour elasticity). Therefore, it is necessary to estimate the production functions separately by type of industrial park for this aspect.

4.2 Analysis results by type

As noted earlier in this section, two types of approaches are taken to examine whether the production factors, relative efficiency and diversity affect production by type of industrial complex. It comes from the background that using industrial park type and location type as a dummy in the overall model would assume that the slope of the remaining independent variables would be the same except for the dummy variables. It may not be right in some sense. Therefore, it is more appropriate to estimate separately for each type, when there is a strong assurance in that there are differences in slope by type of industrial complex and location.

Type 1 is the type of industrial complex, which is divided into national industrial parks and general industrial parks. This is because of the different types of industrial parks, including the size of complexes and tenant companies. Considering another possibility that there may be also differences by type of industrial location, they are divided into four groups: national industrial and general industrial parks in the capital area and national and general industrial parks in the non-capital area. Of course, it can be divided into location types. However, because location types are mixed with each location, the results vary depending on which type of industrial park is included. Therefore, it is decided not to include the analysis in this study because it is difficult to achieve consistency when they are classified by location type.

The relationship between the production factors and the production (see Table 3) showed different results in the influence and order by type of industrial parks. In other words, the overall analysis showed that labour elasticity is greater than capital elasticity. However, according to the type of industrial complex, the national industrial park showed capital elasticity is greater than labour elasticity, and the general industrial park showed that labour elasticity is greater than capital elasticity. It seems likely that the results are affected by the size of general industrial parks, which account for the majority (77.7%) of the 94 old industrial complexes in this analysis. Therefore, it is reasonable to make estimation by the type of industrial park separately in estimating the production function of industrial parks. In addition, there are also differences in the number of production elements when subdividing into location types. But the ranking was found to be the same (see Table 4). In all types of industrial complexes, the capital areas were more resilient than non-capital areas. Therefore, it seems that national industrial parks will need strategies for promoting capital, while general industrial parks will need strategies for promoting employment. Within the same type, complexes in capital areas need capital-oriented strategies and complexes in non-capital areas will need promoting employment-oriented activities.

Next, economy of scale is shown to be different depending on industrial complex type rather than location type. In terms of economy of scale, general industrial parks show economy of scale, while national industrial parks show dis-economy of scale (see Table 3). The same is true for each location type (see Table 4). Therefore, the type of negative economy of scale would require measures to increase output (production) rather than input (capital or employment), while the type of economy of scale would need to seek measures to boost input size rather than production.

In addition, relative efficiency shows positive statistical significance, as in the overall analysis, even though type is subdivided. However, the size of relative efficiency in national industrial parks is larger than that in general industrial parks, and the figures of non-capital areas are higher than those of the capital areas (see Table 3). When it is subdivided by type, the size of relative efficiency was shown in the order of national industrial parks in the non-capital area, national industrial parks in capital areas, general industrial parks in the non-capital areas and general industrial parks in the capital areas (see Table 4). Therefore, it is necessary to focus on general industrial parks and to enable them to achieve efficient operation

Table 3: Production function estimation result (type 1).

	National industrial park				General industrial park			
	Coefficient	Standard error	t value	p value	Coefficient	Standard error	t value	p value
(constant)	9.919	0.329	30.17	0.0001	6.775	0.155	43.62	0.0001
ln area	0.467	0.021	22.02	0.0001	0.343	0.042	8.23	0.0001
ln number of workers	0.380	0.029	12.93	0.0001	0.736	0.033	22.02	0.0001
ln DEA	1.139	0.037	30.90	0.0001	0.783	0.031	25.40	0.0001
ln Herfindahl reciprocal	−0.371	0.093	−4.00	0.0001	−0.058	0.043	−1.33	0.1830
Dummy for capital	−0.207	0.066	−3.16	0.0022	−0.149	0.064	−2.32	0.0212
Modified R^2	0.986				0.953			

Note 1: The reference group in the Seoul metropolitan area is the non-capital area.

of tenant businesses through basic cost-saving strategies and improving the infrastructure of old industrial parks [22–24].

Finally, industry diversity (HHI reciprocal) in type 1 indicates that national industrial parks are statistically significant and robust, whereas general industrial parks have shown a negative sign but have not secured statistical significance (see Table 3).

If the types are further subdivided, statistical significance is not ensured in national complexes in capital areas and general complexes in non-capital areas (see Table 4). This is believed to have occurred due to the small difference in the value of the diversity index of national complexes in the capital areas and general complexes in non-capital areas. However, other types of industrial parks (general industrial parks in the capital areas and national industrial parks in non-capital areas) showed that the higher the industrial diversity in old industrial complexes, the more negative effects on production. This can be interpreted as a kind of integrated economy in which the aggregation of a particular group of industries has more influence on production than the diverse group of industries. However, it was found that for general industrial parks in non-capital areas, the regression coefficients for the diversity of industries are positive or not statistically significant. Therefore, when it comes to the distribution of industries in the regeneration of old industrial complexes in the future, it is deemed to be necessary to consider the linkages with major industries rather than seeking diversity.

Table 4: Production function estimation result (type 2).

		National industrial park				General industrial park			
		Coefficient	Standard error	t value	p value	Coefficient	Standard error	t value	p value
Capital areas	(constant)	10.571	1.342	7.88	0.0001	6.230	0.232	26.86	0.0001
	ln area	0.509	0.077	6.59	0.0001	0.514	0.089	5.75	0.0001
	ln number of workers	0.214	0.092	2.34	0.0305	0.684	0.071	9.64	0.0001
	ln DEA	1.019	0.217	4.69	0.0002	0.587	0.066	8.94	0.0001
	ln Herfindahl reciprocal	−0.098	0.574	−0.17	0.8667	−0.312	0.084	−3.71	0.0004
	Modified R^2	0.977				0.929			
Non-capital areas	(constant)	9.084	0.366	24.85	0.0001	7.204	0.173	41.54	0.0001
	ln area	0.489	0.024	20.41	0.0001	0.281	0.042	6.72	0.0001
	ln number of workers	0.438	0.030	14.80	0.0001	0.730	0.034	21.53	0.0001
	ln DEA	1.088	0.036	30.35	0.0001	0.864	0.032	26.69	0.0001
	ln Herfindahl reciprocal	−0.375	0.086	−4.35	0.0001	0.041	0.046	0.89	0.3721
	Modified R^2	0.990				0.966			

5 CONCLUSIONS AND IMPLICATIONS

This study examined how the relative efficiency and diversity of old industrial parks really affect the production. For this purpose, the study used the industrial park status of the Korea Industrial Complex Corporation and raw data of individual industries as of 2014–2017, based on 94 aged industrial complexes to construct the analysis data. The model is estimated, including relative efficiency and the HHI reciprocal, using the basic model. However, it was approached separately, assuming that the effects of industrial complexes and location types would be different.

The analysis results and implications are as follows. First, the production elements of old industrial parks showed different influence on production by the type of industrial complexes. National industrial parks had a greater influence on the determination of production by capital rather than labour, while general industrial parks had higher contribution to the production by labour than capital. Therefore, it seems that national industrial parks will need promoting capital, while general industrial parks will need promoting employment. Within the same type, capital areas need capital-oriented strategies while non-capital areas will need promoting employment-oriented activities.

Second, it was analysed that national industrial parks enjoy dis-economy of scale and general industrial parks enjoy economy of scale. Therefore, negative economy of scale would require measures to increase output (production) rather than input (capital or employment), while economy of scale would need to seek measures to boost input size rather than production.

Third, the relative efficiency of old industrial parks has a statistically (+) significant effect on production regardless of the type of industrial parks (national and general industrial parks) and the type of locations (capital and non-capital areas). Therefore, the strategies for improving efficiency for old industrial complexes are deemed to have a positive impact on production. In the practical field, it is necessary to take measures to revitalize old industrial parks for the enhancement of relative efficiency by reducing costs of companies operating in industrial complexes. The existing research also suggested measures to secure operating rates, remodel old facilities, increase external effects, diversify industries and improve internal management inefficiency, so the focus should be on this. One of the other ways to reduce costs at the level of old industrial parks is to improve infrastructure. For this reason, national budgets for the regeneration of old industrial complexes should be injected [22–24].

Fourth, the diversity of old industrial parks has basically had negative effects on production as opposed to the result of relative efficiency, while the statistical significance is not secured in national industrial parks in capital areas and general industrial parks in the non-capital areas. The result is viewed to be due to the lack of close examination of the industrial park's connection with the industry when selecting the tenant companies. In particular, closer approach is required, in terms of the diversity of industries in old industrial parks, including industrial links with those that move into the industrial parks in selecting of additional businesses. Therefore, it is necessary to be cautious about strategies to diversify the industry in reviving old industrial complexes. In reality, there is a limit that it is very hard to reorganize the industries of old industrial parks that have already moved in and are in operation. In spite of the problems, it is necessary to introduce highly efficient industrial groups within aged industrial parks or those in the surrounding areas. The outcomes of industrial diversity in this study can be regarded as the diversity in manufacturing industries, since the figures were measured in the middle class of industrial classification for companies in industrial parks. However, it should be noted that the results of this study may be used cautiously, since the recent development of industrial complex tends to contain a mixture of manufacturing and service industries together.

The limitations of this study also exist as always. First of all, inclusion of only the 94 industrial parks in this study is the main limitation so that constrains generalization of the outcomes of relationship among relative efficiency, industry diversity and production. Hence, this research product cannot be interpreted as the nation's entire industrial complex. Access to the data in all industrial parks is necessary for future generalizations, even though it is not easy since it takes a lot of time to organize the related data. Another limitation of the research is that it is assumed that each old industrial complex is a single group. If it is possible, more detailed analysis could be made using detailed information on the companies in each old industrial complex. In spite of all the limitations, we believe that this study is meaningful in attempting to determine the impact of relative efficiency and diversity on old industrial parks that were not covered by previous research.

ACKNOWLEDGEMENTS

This study was conducted with the support of the Ministry of Land, Infrastructure and Transport/the National Institute of Land, Infrastructure and Transport (Task No. 19AUDP-B119346-04).

REFERENCES

[1] Ahn, Y.J. & Lee, M.H., Comparative analyses of efficiency indices in the old industries complex applying data envelopment analysis (DEA) methods. *Journal of The Korean Regional Development Association*, **27(2)**, pp. 219–242, 2015.

[2] Kim, Y.D. & Cho, J.M., A study on the efficiency analysis of industrial complex firms and non-industrial complex firms. *Productivity Review*, **31(4)**, pp. 159–185, 2017.

[3] Choi, M.S., Jang, S.I. & Park, H.Y., Analysis on the relative efficiency of industrial complexes by type based on the DEA model. *Korea Real Estate Review*, **28(3)**, pp. 37–52, 2018.

[4] Choi, J.Y., The effect of fiscal decentralization on economic growth in Korea: focused on production efficiency. *Korean Public Administration Review*, **49(3)**, pp. 161–191, 2015.

[5] Byeon, S.I., Analysis on socioeconomic agglomeration and spatial effect of the ICT manufacture industry in Seoul metropolitan area. *GRI Review*, **13(3)**, pp. 241–264, 2011.

[6] Mo, Y.M. & Kang, M.H., A study on the effect of industrial structure by district in the local economy. *Journal of Korea Regional Economics*, **13(1)**, pp. 81–100, 2015.

[7] Park, S.H., Static and dynamic agglomeration economies in Korea. *Journal of Korea Regional Economics*, **13(1)**, pp. 81–100, 2015.

[8] Kim, Y.S., A study on the determinants of total factor productivity in Korea's regional manufacturing industry. *Journal of Korea Planning Association*, **38(5)**, pp. 199–212, 2003.

[9] Lee, Y.S., Metro cities' and provinces' total factor productivity and its determinants in Korea. *The Korea Spatial Planning Review*, **58**, pp. 39–53, 2008.

[10] Park, D.Y., Seo, B.J. & Jung, C.M., The effect of manufacturing firms' spatial distribution on the productivity of manufacturing industries in SMA. *Journal of Korea Planning Association*, **44(6)**, pp. 147–160, 2009.

[11] Kim, J.H. & Choi, M.S., Economic benefit of renovating deteriorated industrial districts: production function approach. *Journal of the Korea Real Estate Analysts Association*, **23(4)**, pp. 53–63, 2017.

[12] Gujarati, D.N. & Porter, D.C., *Basic Econometrics,* **5th ed.**, McGraw-Hill International Edition, 2009.

[13] Lee, J.D. & Oh, D.H., *Theory of the Efficiency Analysis*, Seoul: Jiphil Media, 2012.

[14] Ryu, S.Y. & Yoon, S.M., Industrial diversification of the wide-economic zones and unemployment rate. *Journal of the Korean Regional Science Association*, **23(3)**, pp. 27–43, 2007.

[15] Ryu, S.Y., Choi, K.H., Ko, S.H. & Yoon, S.M., The impact of industrial diversity to unemployment and employment instability: An analysis of regional economy using panel regression model. *Journal of the Economic Geographical Society of Korea*, **17(1)**, pp. 129–146, 2014.

[16] Moon, D.J. & Hong, J.H., A study on the difference of industrial diversification's impact on local economic growth by city size and location. *The Korean Journal of Local Government Studies*, **19(3)**, pp. 125–152, 2015.

[17] Park, H.Y. & Park, J.H., Analysis of the typology and factors affecting the decline in old industrial parks. *Korea Real Estate Review*, **27(4)**, pp. 7–20, 2017.

[18] Kim, Y.Y., Efficiency analysis of resident companies in industrial cluster complex. *Journal of Economics Studies*, **26(4)**, pp. 157–181, 2008.

[19] Lho, S.W., Analysis on the efficiency and productivity for machinery firms located at Changwon using DEA and Malmquist method. *Journal of Economics Studies*, **32(4)**, pp. 237–260, 2014.

[20] Lee, Y. & Ahn, Y.H., Analysing efficiency of the selected national industrial complexes in Korea using DEA and Malmquist productivity index. *Journal of The Korean Regional Development Association*, **23(5)**, pp. 95–118, 2011.

[21] Kim, S.H., Choi, M.S. & Kim, E.J., Temporal spatial externalities on agglomeration economy of manufacturing : estimation of spatial SUR by using 3SLS. *Journal of the Economic Geographical Society of Korea*, **10(4)**, pp. 414–426, 2007.

[22] Korea Development Institute, *Daegu 3rd Industrial Complex Regeneration Project*, 2011a.

[23] Korea Development Institute, *Daejeon Industrial Complex Regeneration Project*, 2011b.

[24] Korea Development Institute, *The West Daegu Industrial Complex Regeneration Project*, 2011c.

HOW DO CITIES OF DIFFERENT SIZES IN EUROPE WORK WITH SUSTAINABLE DEVELOPMENT?

ANNA SÖRENSSON, MARIA BOGREN & ULRICH SCHMUDDE
ETOUR, Mid Sweden University, Sweden.

ABSTRACT
Today, competition between cities to attract inhabitants, companies and tourists is strong and cities must be up-to-date in terms of development to succeed. One way for smaller destinations to achieve sustainable development is by being creative with respect to tourism. Some destinations are in the shadow of others and need to stand out in some way. The purpose of this paper is to study European cities of different sizes and their work on sustainable development. The following research questions are addressed: How do different cities work on sustainable development? How does the size of the city influence its work on sustainability? How can a city be influenced by surrounding areas in its sustainability work? The study uses a qualitative method. Data were collected from 34 small cities, towns and villages in Europe. The destinations were selected using non-probability sampling. The data were analysed using an interpretative approach. The results show that the local community plays a key role in contributing to the sustainable development of small destinations. It is also of great importance for a place to have an identity and to reach different types of stakeholders. The conclusion is that local communities must be engaged in the sustainable development of smaller destinations. It is also important to focus on the environment since today's tourists are more aware of environmental sustainability. Several of the cities can be seen as shadow destinations since they are included in larger regions and are dependent on other destinations that are more famous. Finally, successful rural destinations offer value to the tourist, have a strong identity and include stakeholders in the development process.
Keywords: creativity, identity, rural, small city, sustainable, sustainable development.

1 INTRODUCTION
There is strong competition today between cities to attract inhabitants, companies and tourists, and destinations must be up-to-date in their development. This is particularly difficult for smaller villages and rural towns. The focus of this study is cities, towns and villages. One way for smaller destinations to achieve sustainable development is by being creative and exploiting tourism niches [1]. Sustainable development is defined in many ways, but the most frequently quoted definition is from the 1987 Brundtland Report [2], which states, *'Sustainable development is development that meets the needs of the present without compromising the ability of future generations to meet their own needs'*. Another dimension of sustainable development is that it consists of three pillars, namely economic, social and environmental sustainability. Middleton and Clarke [3] identify a need to find a balance between these three dimensions of sustainability. Today, the concept of sustainable development is found in all types of industries and research fields, but there is still a contradiction between development and sustainability [4]. It is challenging to achieve both economic growth and environmental protection in cities.

Sustainable development is an important issue for cities in Europe and around the world. Cities have to take care of their inhabitants as well as attracting tourists. Previous research has shown that the principle of sustainability is widely recognised but its implementation is more limited [5] [6] [7] [8]. Companies at the destination may be willing to apply the concept of sustainability for their own benefit as long as it increases their revenue and improve their public relations and does not cost them much. Sustainability can often be used as a marketing strategy, a phenomenon known as greenwashing [9]. The concept of sustainable development

© 2020 WIT Press, www.witpress.com
DOI: 10.2495/DNE-V14-N4-287–298

is trendy and can create goodwill for a company and even a whole destination. Investing in energy-saving measures and water reduction systems can also enable a company to save money. Creativity is important for sustainable cities, and one way to gain economic growth is through tourism. Richards [10] argues that tourism nowadays is creative in many ways. Creative tourism influences a place's identity, shaping what the place stands for in the minds of both tourists and inhabitants.

Research has shown that different countries define cities differently based on a wide range of criteria. *'These criteria often include population size and density, but also more functional or historic ones such as having urban functions, being a recipient of national urban policy funds or having received city rights through a charter sometime between the Middle Ages and today. Comparing the number of cities based on national definitions across countries is hopelessly distorted by difference in methodology* [11].'

This study defines cities as follows:

1. Village – a village is a human settlement or community that is larger than a hamlet but smaller than a town (0–999 inhabitants).
2. Town – a town has a population of 1,000 to 9,999.
3. City – a city has a population of 10,000 to 50,000.

This paper uses the words 'place', 'destination' or 'city' when the sense is general and employs the above terms when size (village, town or city) is relevant. A particular type of destination has received little attention from researchers. These destinations are situated close to high-profile tourist attractions (such as cities) but do not always benefit fully from this proximity. Hudman and Jackson [12] argue that destinations that are situated near popular attractions may be affected by the shadow effect: *'The shadow effect refers to destinations that are near other major destinations. The concept comes from a geographic term 'rain shadow'. Some localities get less rain because the precipitation is diverted by mountains or wind patterns. Thus, one destination may be in the shadow of another destination, which is the preferred destination. Because they are close to the preferred destination, tourists will also visit the shadow destinations, but stay less time'* [12]. Tourists that visit a certain attraction might also want to experience other attractions in the nearby region. Tourists often see a destination as one unit, despite the fact that it is a complex network that involves a large number of actors [13]. Another problem is how to determine a destination geographically, i.e. where a destination starts and finishes, particularly in the minds of tourists. The importance of treating a destination as a unit has resulted in a large amount of research focusing on issues related to destination development [13]. Destinations that are situated geographically close to a famous destination can benefit from the strong brand of the well-known destination and use this to increase its own tourist numbers [14]. Ashton [15] states that *'brand is considered as a powerful instrument in creating a successful destination'*. Destinations with strong brands have a clear identity. Tourists recognise them and, therefore, experience a feeling of familiarity. Shadow destinations can, therefore, develop relationships with famous destinations close-by and benefit from their branding. Research has not yet addressed the issue of sustainable development of cities to any great extent. The main focus is often on companies and their work on sustainability. By contrast, this study focuses on the sustainable development of cities. The purpose of this paper is to study European cities of different sizes and their work on sustainable development.

RQ1: How do different cities work on sustainable development?

RQ2: How does the size of the city influence its work on sustainability?

RQ3: How can a city be influenced by surrounding areas in its sustainability work?

2 THEORETICAL FRAMEWORK

The theoretical framework is structured around three areas, namely (1) sustainable development, (2) creativity and (3) identity.

2.1 Sustainable development

As previously mentioned, the starting point of sustainable development was the 1987 Brundtland Commission Report [2]. The issue is often discussed in relation to the three pillars of economic, social and environmental sustainability (see Fig. 1 below). The idea is that these three areas are tied together within the concept of sustainable development. However, previous studies have shown that some destinations focus more on one of the dimensions rather than focusing equally on all three [4].

Since the Brundtland Commission Report, work on sustainable development has continued through measures such as the Agenda 21, Rio Declaration and the recent establishment of 17 sustainable development goals (SDGs): *'The 2030 Agenda for Sustainable Development, adopted by all United Nations Member States in 2015, provides a shared blueprint for peace and prosperity for people and the planet, now and into the future. At its heart are the 17 Sustainable Development Goals (SDGs), which are an urgent call for action by all countries – developed and developing – in a global partnership. They recognise that ending poverty and other deprivations must go hand-in-hand with strategies that improve health and education, reduce inequality, and spur economic growth – all while tackling climate change and working to preserve our oceans and forests'* [17].

The main obstacle for sustainable development is how to operationalise it at a practical level [18]. Sustainable development as a concept has developed in several ways since 1987.

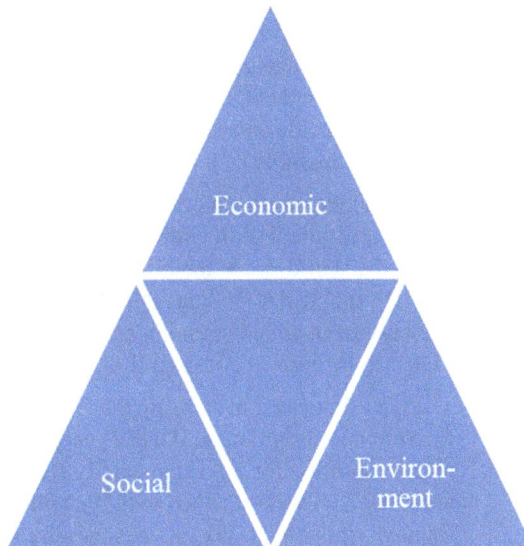

Figure 1: Dimensions of sustainability.
Source: [16].

Lam and Yap [19] identify several sustainability frameworks in the literature (see Table 1 below). These different frameworks can be seen as structures to study sustainable development. They can also be helpful in structuring information for reporting and communication. The first framework is defined as 'driving force-state-response' and refers to the pressure created by society through the implementation of policies to control the impact of human activities on the environment [19]. The second framework is labelled the 'theme framework'

Table 1: Existing Sustainability Frameworks.

Framework	Description
Driving Force-State-Response (DSR)/ Pressure State-Response (PSR) Framework	Human activities exert pressure on the environment, causing changes in the state of the environment or the quality and quantity of resources through emissions and consumption. Society then responds to these changes by instituting policies, which, in turn, mitigate the pressure. It was adopted by the European Environmental Agency (EEA) and the European Statistical Office in 1997.
Theme Framework	The thematic framework compiles a comprehensive list of indicators concerning various themes or issues related to sustainability. It facilitates the monitoring of progress towards goals and is flexible so that indicator sets can be adjusted to new policies.
Capital Framework	The capital framework calculates national wealth as a function of the sum of and interactions among different kinds of capital, including financial capital, capital goods produced and natural, human, social and institutional capital expressed in monetary terms.
Systems Analytical Framework	Within the systems analytical framework, sustainable development indicators (SDIs) are chosen based on their ability to provide answers to a set of questions with regard to the sustainability of a system.
System of Integrated Environmental and Economic Accounting (SEEA) Framework	Pioneered by various international bodies (the United Nations Statistical Commission with the International Monetary Fund, the World Bank, the European Commission and the OECD), the SEEA facilitates the construction of a common database from which common economic and environmental SDIs can be derived consistently.
Global Reporting Initiative (GRI)	The GRI (launched in 1997 by the United Nations Environment Programme [UNEP] and the United States NGO, Coalition for Environmentally Responsible Economics [CERES]) was launched with the aim of enhancing the 'quality, rigor and utility of sustainability reporting.' The GRI uses a hierarchical framework in three areas—social, economic and environmental.

Source: [19].

and relates to economic, social and environmental dimensions. These three dimensions are also used by the Global Reporting Initiative (GRI), which is responsible for the sixth framework listed in the table. Sustainability reporting can be conducted in many ways but the most frequently used guidelines for sustainability reporting are those of the GRI. The GRI is a non-governmental organization (NGO) founded in Boston in 1997 with the aim of helping companies to produce sustainability reports (i.e. in relation to the economic, social and environmental dimensions). Many other guidelines exist, but research has shown that the GRI's guidelines are the most commonly used in many countries [20]. They is, therefore, regarded as the current standard. The third framework is the 'capital framework,' which is based on monetary terms. From the perspective of this framework, sustainability can be calculated. However, some researchers claim that expressing capital in monetary terms has limitations. The fourth framework is the 'systems analytical framework,' which refers to dependence on the system. According to this framework, sustainable development is based on the system's ability. The fifth framework is the 'System of Integrated Environmental and Economic Accounting (SEEA) framework,' which was established by the United Nations to collect sustainability information consistently in a database [19].

This table shows the numerous frameworks relating to sustainable development. The most commonly used framework is based on the three dimensions (economic, social and environment). Nowadays, cities are becoming more creative in terms of sustainable development.

2.2 Creativity

Creative solutions can help destinations to develop sustainably. Research on cities has shown that creativity plays a role in development [10]. Nowadays, it is a common perception that creative places attract creative people. These ideas are also entering the field of destination development. Creativity can contribute to tourism and destination development in a number of ways by adding atmosphere to a place, developing tourism products and providing economic spin-offs for creative development. Ashworth and Page [21] argue that there is a paradox in the relationship between tourism and tourist destinations, namely the more attractive a destination becomes, the more its inhabitants are influenced.

Richard [10] argues that creativity manifests in three ways, namely creative industries, creative cities and creative class (see Table 2 below).

Creative cities often focus on capital and on the social dimension of sustainability. For creative cities, tourism has become a competitive strategy to stimulate growth [22]. The cities need to identify opportunities and be prepared to act. A focus on sustainability issues can be a competitive advantage. Growing competition between destinations has led to the development of new themes and branding strategies. Nowadays, destinations must have a strong brand and

Table 2: Key conceptual approaches to creativity in cities.

	Creative industries	**Creative cities**	**Creative class**
Focus	Creative production	Creative milieu	Creative consumption
Form of capital	Economic	Social and Cultural	Creative
Creative content	Arts, media, film, design, architecture, etc.	Creative places, artistic production	Atmosphere and 'cool'

Source: [10].

an identity that attract tourists. Creativity brings a number of advantages for destination branding in terms of targeting both tourists and people who want to live in the area [10].

2.3 Identity

Nowadays, the importance of branding for successful tourism destinations is well known to researchers [23] [24] [25]. Although researchers have not yet accepted a common definition, it is argued that 'brand identity development is a theoretical concept best understood from the supply-side perspective' [26]. Kapferer [27] argues that brand identity is not just a supply-side factor and states, '*before knowing how we are perceived, we must know who we are*'. According to his statement, the tourist destination, rather than the tourists, should define both the destination's brand and character.

Aaker and Joachimsthaler [28] discuss the term brand identity and state that the term specifies what the brand wants to represent to its main target audience and that brand identity acts in many ways. A brand strategist comes up with ideas about what the brand should be associated with and find ways of creating these perceptions. The brand identity also represents a vision of what associations the brand should project to its target audience. This builds a relationship between a brand and its target audience. The brand identity helps the brand to become a valued choice by having its audience associate the brand with benefits and credibility. It is important for a shadow destination to create its own unique brand. A shadow destination can benefit from a neighbouring key destination but still need to determine and build its own brand. When creating and building up a brand, it is important to consider factors such as brand identity, brand image and brand personality. With regard to tourist destinations, Ashton [15] states that it is important to focus on the iconic features of a destination. The brand image presents and establishes these features in the mind of the tourist. When branding a destination, the environment and resources are the basis for branding. In order to create a popular destination, having a brand identity and brand image is important [29] [30]. When creating a strong brand, it is important for a destination to generate brand loyalty among tourists.

3 METHOD

This study takes a qualitative approach involving 34 cases. The cases were selected using non-probability sampling on the basis that they were peripheral, largely rural destinations in Europe with tourist attractions. These selection criteria were chosen due to limited earlier research on sustainability with these kinds of cases. The initial data were collected with the help of bachelor students under the supervision of the authors. Additional data were also collected from secondary sources, including web pages and written materials. As previously explained, this study divides places into three groups: villages (up to 999 inhabitants), towns (with a population of 1,000 to 9,999), and cities (with a population of 10,000 to 100,000). These three groups were used for the data collection. In the dataset, the largest city has a population of around 32,000. The data set consists of 7 villages, 12 towns and 15 cities, including 16 destinations in Sweden, 8 destinations in Germany and 10 destinations in other European countries. For an overview of the destinations, see Table 3 below.

For most of the studied destinations, the main attraction is related to nature or history. Four of the destinations are UNESCO World Heritage Sites. The data were analysed using an interpretative approach based on the three sustainability dimensions (based on the 'theme framework' presented in Table 1). The concepts of creativity and identity are included in sustainability.

Table 3: Description of the selected villages, towns and cities.

Village/Town/City	Country	Inhabit-ants	Village/Town/City	Main attraction
Arjeplog	Sweden	1785	Town	Nature & Car testing
Arvika	Sweden	14023	City	History & Art
Berg	Sweden	1281	Village	Canal & History
Blieskastel	Germany	20770	City	Nature & UNESCO
Bodenmais	Germany	3459	Town	Nature & Ski
Droux	France	420	Village	Hyperloop Tunnel
Finspång	Sweden	13279	City	History
Gällivare	Sweden	10329	City	Nature
Gällö	Sweden	722	Village	Camping & Ski tunnel
Hattem	Netherlands	12108	City	History & Museum
Lescar	France	10393	City	History
Lienz	Austria	11844	City	History & Nature
Lycksele	Sweden	8572	Town	Zoo & Nature
Maratea	Italy	12108	City	Sea & Nature
Meckenheim	Germany	24661	City	History & Nature
Motala	Sweden	31385	City	Canal & History
Mörsdorf	Germany	581	Village	History & Nature
Norrtälje	Sweden	20721	City	Sea & History
Nynäshamn	Sweden	14864	City	Sea & Port City
Parkstetten	Germany	3158	Town	Nature & History
Porto Venere	Italy	4041	Town	Sea & UNESCO
San Gimignano	Italy	7774	Town	History & UNESCO
Sandhornoya	Norway	345	Village	Nature
Sankt Englmar	Germany	1861	Town	Nature & Religion
Sankt Goarshausen	Germany	1280	Town	Nature & UNESCO
Sassnitz	Germany	9435	Town	Sea & Port Town
Stora blåsjön	Sweden	200	Village	Nature
Storlien	Sweden	74	Village	Nature & Shopping
Trelleborg	Sweden	30436	City	Sea & Port City
Vadstena	Sweden	5764	Town	Religion & History
Valle del Jerte	Spain	11123	City	Spa & Nature
Visby	Sweden	24272	City	History & Sea
Vuokatti	Finland	6183	Town	Nature & Ski
Åkersjön	Sweden	100	Village	Nature & Snowmobiling

Source: own creation

4 RESULTS AND DISCUSSION

In this part, we use the term place or destination for all three types (i.e. village, town or city). It is only when we write about a certain group that we use these three subgroups.

4.1 Economic sustainability and identity

For all of the studied destinations, attracting tourists is important to earn income. Villages must have functioning businesses to make money. Economic sustainability is about revenue, market shares and indirect economic impact. If people want to live in a certain place, they must be able to make money somehow. In cities, there is often more choice for inhabitants with regard to finding a job or running a business. In smaller towns or villages, there are fewer options. Therefore, small villages are more dependent on visitors. However, there are so many tourist destinations, so competition among European cities is tough. Therefore, it is beneficial to create an identity and a brand that sticks in the minds of tourists and gives them a reason to travel to the destination. An identity makes a destination known for something. Nowadays, both visitors and inhabitants look for creative environments. Creative cities are often focused on social and cultural dimensions, but there is first a need for economic sustainability.

All of the destinations studied seem to work on different layers of sustainability. All of them want tax revenues and income from visitors. When they have some sort of economic sustainability they are able to focus on the next step, which is the social dimension. The larger cities have achieved economic sustainability and are, therefore, also able to work on the social and environmental dimensions of sustainability. These dimensions are discussed below. Many of the destinations studied are part of larger destinations at the regional or county level. They are, therefore, in the shadow of more famous destinations or dependent on larger regional destinations. For example, the town of Bodenmais is part of the Bavarian Forest, which is a more famous destination with a stronger identity. Four of the destinations have a strong identity as they are part of the UNESCO network. Research has shown that the UNESCO label is a well-established brand that can be used in to communicate a strong identity.

4.2 Social sustainability and network

Regarding social sustainability, the local community is a key actor for a destination's survival. From the destination's perspective, the social dimension starts with the inhabitants. All places have a local culture, which is strongly connected with the inhabitants. Visitors want to experience the local culture, and, if too many visitors come, this can influence the local culture. Many of the cities studied, and some of the towns, have policies on social issues in their sustainability plans.

The results show that cooperation is often found within a destination, where different types of service providers work together and build networks. It is often the main actors at the destination that cooperate. The results also show that villages are more aware of the importance of cooperation. As they are small communities, it is necessary to cooperate to be able to meet the needs of tourists. The role of the community is a key success factor for the sustainable development of villages, towns and cities. One example is Sankt Englmar, where there is significant cooperation between hotel owners, the mayor and activity companies. Voukatti in Finland is another example of a destination where there is strong cooperation between politicians, companies and sports clubs. They focus on maintaining existing relationships rather than reaching new visitors. The result is loyal customers who come back year after year.

The study indicates that smaller villages see each other as competitors as well as partners. Their mentality is to support each other, and they are highly dependent on each other. Another finding is that the villages regard themselves as part of larger communities. It can be a tourism region or a brand of a region. Åre is the largest ski resort in Sweden and is situated around 61 km from Storlien. When tourist in Åre visit the web page for Åre [31], information about Storlien can be found. This is one example of a village (Storlien) that benefits from belonging to a larger 'social' community.

4.3 Environmental sustainability and geography

Environmental sustainability is less of a focus among smaller villages, even though their main attraction is nature. As shown in Fig. 2, the environment is the third and last sustainability dimension that destinations focus on. Many of the destinations studied, regardless of their size, are most famous for a natural attraction. For these places, nature is the main attraction for tourists. The prerequisites for nature attraction vary depending on the places location, for example, if they are located by the sea or inland by the mountains. This is also something that destinations can use to strengthen their identity. For example, Bodenmais highlights its unspoiled nature as it has one of the last two 'jungles' in Europe. Sustaining the environment is important; inhabitants and visitors are no longer accepting of pollution and unnecessary waste. Inhabitants are often more aware of the environment in their home than when visiting a destination. However, visitors will not visit a polluted city. A destination can benefit economically from focusing on environmental sustainability measures, such as energy and water consumption. Lower consumption lowers a city's costs. The results are summarised in the following Table 4 below.

In summary, the larger the city, the more they formal its sustainability plans. The economic dimension of sustainability is a foundation for working on the social dimension and, lastly, the environmental dimension (see Fig. 2). Networking and creativity and establishing a strong identity are more important for villages to prevent them from becoming shadow destinations. The dependence of the ones in their geographical area influences their sustainability work.

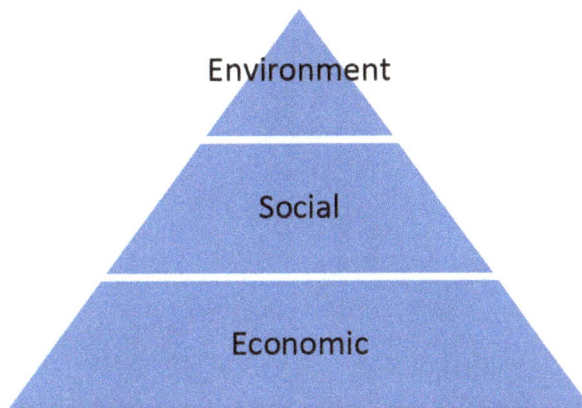

Figure 2: The layers of sustainability in cities.
Source: own creation.

Table 4: Different types of destinations and sustainable development.

	Sustainable Development of the City
Villages	• Do not have sustainable development to any larger extent in public written form • The community of great importance • Are often 'shadow destinations' and are therefore part of larger destinations, areas or regions
Towns	• Often focus on social sustainability, whereby the inhabitants and their wellbeing are prioritised • Are eager to satisfy their inhabitants in order to gain tax revenues • When towns develop, they focus on the environment
Cities	• Have formal and structured sustainability programmes and documented policies • Are often the capital city in the municipality • Environmental sustainability is often strongly linked to the local environment (e.g. cities by a lake or sea focus on water quality)

Source: own creation.

4.4 Theoretical contribution

This study has generated a model that can be used by shadow destinations for sustainable development. This builds on a model developed by Bogren *et al.* [16]. It incorporates not only the three dimensions of sustainable development but also includes factors that influence the sustainable development of shadow destinations (e.g. identity, network and geography).

The interaction between a place's identity and its network are addressed using the Destination Management Organisation (DMO). The findings of the study show that DMO plays an important role in promoting a region or area. The results also show that identity and geographical region play key roles. As previous research has shown, tourists may see a certain area as a destination even though, in reality, it may comprise two different cities or areas. The identity of

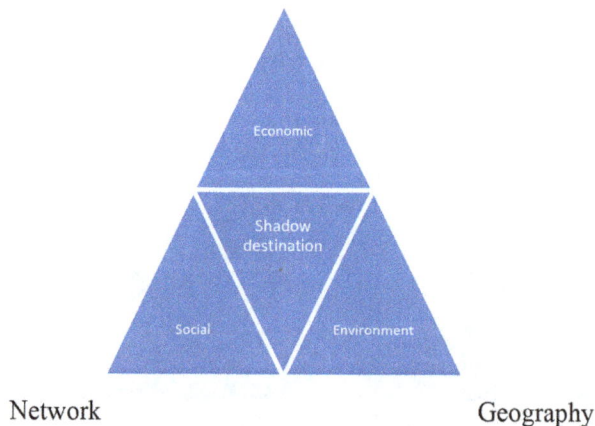

Network   Geography

Figure 3: Key Success Factors for Sustainable City Development.

Source: own creation based on [16].

a place (e.g. destination) plays a key role in relation to the geographical area. This is particularly interesting for destinations that are in the shadow of a more well-known destination with a strong identity. The third relation is between network and geography. The results show that the local community of a destination plays a key role in sustainable development. Another aspect of great importance is infrastructure, i.e. how easy the destination is for tourists to access. This model can be used by shadow destinations in their work towards sustainable development.

5 CONCLUSIONS

The main conclusion is that the cities work on sustainable development in a more structured way than villages and towns. It is also found that the size of a destination influences its work on sustainable development. Smaller villages do not carry out any sustainability work of their own but rather depend on nearby towns, cities or regions (often the municipality's capital). In this regard, they can be seen as shadow destinations, and their sustainable development is dependent on others. The conclusion is that the local community must be engaged in sustainable development of smaller rural areas. It is also important to focus on the environment since today's tourists are more aware of environmental sustainability. Finally, successful rural destinations offer value to tourists, have a strong identity and include stakeholders in the development process.

REFERENCES

[1] Carlisle, S., Kunc, M., Jones, E. & Tiffin, S., Supporting innovation for tourism development through multi-stakeholder approaches: Experiences from Africa. *Tourism Management*, **35**, pp. 59–69, 2013. https://doi.org/10.1016/j.tourman.2012.05.010

[2] Brundtland Commission Report, 1987, https://sustainabledevelopment.un.org/content/documents/5987our-common-future.pdf, (accessed 23 June 2019)

[3] Middleton, V. & Clarke, J., *Marketing in Travel and Tourism*, Jordan Hill, 2001.

[4] Sörensson, A. Sustainable mass tourism: Fantasy or reality?. *International Journal of Environmental, Cultural, Economic and Social Sustainability*, **7(5)**, pp. 325–334, 2011. https://doi.org/10.18848/1832-2077/cgp/v07i05/54998

[5] Liu, Z., Sustainable tourism development: A critique. *Journal of Sustainable Tourism*, **11(6)**, pp. 459–475, 2003. https://doi.org/10.18848/1832-2077/cgp/v07i05/54998

[6] Sharpley, R., Tourism and sustainable development: Exploring the theoretical divide. *Journal of Sustainable Tourism*, **8(1)**, pp. 1–19, 2000. https://doi.org/10.1080/09669580008667346

[7] Saarinen, J., Traditions of sustainability in tourism studies. *Annals of Tourism Research*, **33(4)**, pp. 1121–1140, 2006. https://doi.org/10.1016/j.annals.2006.06.007

[8] Wall, G., Sustainable development: Political rhetoric or analytical construct?. *Tourism Recreation Research*, **27(3)**, pp. 89–91, 2002. https://doi.org/10.1080/02508281.2002.11081377

[9] Delmas, M.A. & Burbano, V.C., The drivers of greenwashing. *California Management Review*, **54(1)**, pp. 64–87, 2011. https://doi.org/10.1525/cmr.2011.54.1.64

[10] Richards, G., Creativity and tourism in the city. *Current Issues in Tourism*, **17(2)**, pp. 119–144, 2014. https://doi.org/10.1080/13683500.2013.783794

[11] European Parliament, https://ec.europa.eu/regional_policy/sources/docgener/focus/2012_01_city.pdf, (accessed 23 June 2019)

[12] Hudman, L.E. and Jackson, R.H., *Geography of Travel & Tourism*. Cengage Learning, 2003.

[13] Haugland, S.A., Ness, H., Grønseth, B.O. & Aarstad, J., Development of tourism destinations: An integrated multilevel perspective. *Annals of Tourism Research*, **38(1)**, pp. 268–290, 2011. https://doi.org/10.1016/j.annals.2010.08.008

[14] Schmudde, U. & Sörensson, A., *Tourism Development in Rural Areas in Sweden—In the Shadow of a Well-Established Destination*, Aten: Atiner, 2019.

[15] Ashton, A.S., Tourist destination brand image development—an analysis based on stakeholders' perception: A case study from Southland, New Zealand. *Journal of Vacation Marketing*, **20(3)**, pp. 279–292, 2014. https://doi.org/10.1177/1356766713518061

[16] Bogren, M., Cawthorn, A. & Sörensson, A., Sustainability among large-sized companies in Europe—are there national differences in their sustainability information? ed. I. Bernhard, Uddevalla Symposium 2018: Diversity, Innovation, Entrepreneurship—Regional, Urban, National and International Perspectives, 2018. https://symposium.hv.se/globalassets/dokument/forska/symposium/content-list-proceedings-2018, (accessed 23 June 2019)

[17] UN. https://sustainabledevelopment.un.org/?menu=1300, (accessed 23 June 2019).

[18] Pavlovskaia, E., Sustainability criteria: their indicators, control, and monitoring (with examples from the biofuel sector). *Environmental Sciences Europe*, **26(1)**, pp. 1–16, 2014. https://doi.org/10.1186/s12302-014-0017-2

[19] Lam, J.S.L. & Yap, W.Y.A., Stakeholder perspective of port city sustainable development. *Sustainability*, **11(447)**, pp. 1–15, 2009. https://doi.org/10.3390/su11020447

[20] Kuzey, C. & Uyar, A., Determinants of sustainability reporting and its impact on firm value: Evidence from the emerging market of Turkey. *Journal of Cleaner Production*, **143**, pp. 27–39, 2017. https://doi.org/10.1016/j.jclepro.2016.12.153

[21] Ashworth, G. & Page, S.J., Urban tourism research: Recent progress and current paradoxes. *Tourism Management*, **32(1)**, pp. 1–15, 2011. https://doi.org/10.1016/j.tourman.2010.02.002

[22] Stolarick, K.M., Denstedt, M., Donald, B. & Spencer, G.M., Creativity, tourism and economic development in a rural context: The case of Prince Edward County. *Journal of Rural and Community Development*, **5(1)**, 2011.

[23] Ashworth, G. & Kavaratzis, M., Beyond the logo: Brand management for cities. *Journal of Brand Management*, **16(8)**, pp. 520–531, 2009. https://doi.org/10.1057/palgrave.bm.2550133

[24] Govers, R. & Go, F., *Glocal, Virtual and Physical Identities, Constructed, Imagined and Experienced*, Springer, 2009.

[25] Kladou, S., Kavaratzis, M., Rigopoulou, I. & Salonika, E., The role of brand elements in destination branding. *Journal of Destination Marketing & Management*, **6(4)**, pp. 426–435, 2017. https://doi.org/10.1016/j.jdmm.2016.06.011

[26] Konecnik, M. & Go, F., Tourism destination brand identity: The case of Slovenia. *Journal of Brand Management*, **15(3)**, pp. 177–189, 2008. https://doi.org/10.1057/palgrave.bm.2550114

[27] Kapferer, J.N., Why are we seduced by luxury brands?. *Journal of Brand Management*, **6(1)**, pp. 44–49, 1998.

[28] Aaker, D.A. & Joachimsthaler, E., The brand relationship spectrum: The key to the brand architecture challenge. *California Management Review*, **42(4)**, pp. 8–23, 2000. https://doi.org/10.1177/000812560004200401

[29] Aaker, D.A., *Brand Relevance: Making Competitors Irrelevant*, John Wiley & Sons, 2010.

[30] Qu, H., Kim, L.H. & Im, H.H., A model of destination branding: Integrating the concepts of the branding and destination image. *Tourism Management*, **32(3)**, pp. 465–476, 2011. https://doi.org/10.1016/j.tourman.2010.03.014

[31] Skistar Åre, https://www.visitare.com/storlien, (accessed 23 June 2019)

REFORMULATING A SMART HOME SYSTEM FOR THE INDIAN CONTEXT: DIU ISLAND

DARPAN TRIBOAN[1] & ANISHA MEGGI[2]
[1] Context, Intelligence and Interaction Research Group, De Montfort University, UK.
[2] Leicester School of Architecture, De Montfort University, UK.

ABSTRACT

In a fast urbanizing world, the Smart City concept driven by leading technologies can be a saviour to the many urban, environmental and economic issues among other problems being faced by governments and citizens. The smart city concept is discussed in conjunction with long explored urban and archi-tectural theories of utopia and ideal city design. Further expanding the conversation on the role of the home as a tool by which to live life, which has been understood as the current concept of Smart Homes (SH) and Ambient Assistive Living (AAL) systems. This paper focuses on a recently announced Smart city in India, Diu Island, which is the primary case study in the paper. Within the context of Diu Island, the issues faced by the elderly native population are investigated to propose a smart home living system that can help improve the quality of their daily lives. A hybrid approach is proposed that leverages edge and cloud computing paradigms to become self-powering, energy-efficient and reduces delays in aiding the elderly.
Keywords: Ambient Assisted Living, Diu Island, Smart City Mission of India, socio-cultural aware, Smart Homes, urbanisation.

1 INTRODUCTION

Smart cities are 21st century technological utopias. When such utopian concepts are applied and practised by economically developing and emerging countries, it creates a complex con-textual environment of aspirations and realities, converging histories and socio-cultural dynamics, [1]. In India, as urbanisation changes family structures and lifestyles, a society deeply rooted in joint family dynamics is becoming fragmented and leaving the elderly at risk of being isolated, neglected within the city. The case study in this paper investigates a recently announced Smart city, Diu Island, where historical, political aspects mean there is high migration of the native population, leaving behind the elderly in many cases. Within this environment of change, the elderly native population on the island face many city-level but first and foremost daily household challenges, which this paper will focus on.

To understand the Smart City concept wholly, the literature review in this paper delves into the architectural and urban theories on ideal cities that capture ideas from the Renaissance ideal cities of the 15–16th centuries. By gaining an understanding of the historical and more contemporary ideologies of managing, organising and designing a city the future of smart cities can also be better evaluated. The smart city concept in the early 20th century can be compared to the technologically based utopian visions of Le Corbusier [2]. His critically debated quote 'house is a machine for living' becomes a pivotal point allowing the research to take on the Smart Home system and specifically the Ambient Assistive Living (AAL) approach to be developed for contextually complex environments.

This paper aims to investigate and re-formulate a smart home system for a culturally com-plex context within Diu Island. The main objectives being; (1) to comprehensively document, critically analyse and evaluate literature in the smart home, smart city environment in India, (2) to select and investigate a problem context that requires smart home interventions, this will be the context of Diu Island specifically the migrant village of Fudam, (3) to recognise

DOI: 10.2495/DNE-V14-N4-299–310

and evaluate methods of smart home interventions for the problem context, (4) to reformulate a method of smart home intervention for the problem recognised in the case study, (5) to evaluate the impending potential and scope of the method applied and propose future avenues of research.

2 RESEARCH METHODOLOGY

This paper utilises a literature review and qualitative primary case study to understand a problem context, which determines the research objectives. The research is constructed in the following stages; awareness, suggestion, development, evaluation, and conclusion, [3].

The literature review covers (1) urbanisation in South Asia, concept of 'Smart Cities', (2) contextualisation of the environment surrounding old age homes and elderly living alone in their homes in India, and (3) various smart home systems available, allow for the recognition and critical comparison and analysis of smart home systems.

The qualitative primary case study of Diu Island will allow for an in-depth contextual investigation of a specific case where contextual complexities will be understood from a wider variety of perspectives. The context will be studied for the recognition of a specific problem area that exists and a smart home system to be applied and evaluated. This will allow for the proposal of a hybrid smart home system that is appropriate for the problems recognised in the case study to apply, test and evaluate the smart home system method.

The primary case study is a part of research conducted in [4], [5] whilst understanding the derelict and decaying built environment, the issues of the elderly inhabitants who had chosen not to leave their homes was recognised. The researcher's site visits and observations of the elderly residents and their lives become an integral part of composing the problem context for this paper. The criteria for analysing this case study are the; political, economic, social, technological, environment and architectural aspects that will be analysed for opportunities and restrictions. The smart home system that is modified and proposed for the problem context recognised in the primary case study will be discussed, evaluated and further research will be proposed.

3 LITERATURE REVIEW

3.1 Smart Cities, Utopia and the future of cities in India

There is no common meaning of a 'smart city' is however all meanings converge on being 'technologically based', [6]. The World Bank suggests two meanings; where there are sensors everywhere, collecting real-time data from interconnected devices, and a city that cultivates better relationships between citizens and the government. It was IBM who coined the 'smart city' phrase in 2008 and is the market leader in smart city suppliers, [7]. There has been some concern due to the roles private corporations are playing in defining what a smart city is, where the smart city is becoming more of a corporate smart city. A case study on the smart city of Genoa exemplifies how the smart city concept acts as a promoter of interests for the business elites diverting the attention away from urgent problems such as urbanization, [6].

In the Smart Cities Mission for India, the government body recognises there is no set meaning for the term smart city hence, smart city would have a different connotation in India than it has in Europe, [7]. An example of the failure of a smart city whose model was implanted straight from Europe into India is Lavasa, [8]. The overall objective of smart growth anywhere would be for sustainable, economic and social progress to occur, [9].

Smart City initiatives are said to be agreed between the government and private companies, which questions the marginalisation of vulnerable populations. It has been reported that the smart city initiatives in India have meant forced acquisition of land and relocation of street vendors and middle class communities from the centre of cities to other areas, [7]. Some specific recommendations in smart city initiatives is citizen participation that should be integral to all parts of decision making, [7]. Within smart city initiatives, the idea that one size fits all is most problematic as each city being designated the smart city title has its own identity, cultural values and ways of life which need to be sustained in its process to becoming 'smarter'. In addition, this can only be done when the citizens who are key components of the city's identity are part of the different levels of decision making and implementing and testing.

Borja argues that, technology does not always play in the favour of citizens, within the smart city where technology is one of the basic principles. In 2013 Greenfield in, 'Against the Smart Cities', states that the concept of the intelligent, smart city has been completely derived by private companies. The smart city is stated to be an ideological technological utopia for the future of cities, [9]. More specifically, concrete utopia distinguished by E. Bloch, 1995,refers to a project connected to reality allowing citizens to progress towards historical and social transformation.

Utopian thinking has played a critical role in the discipline of city planning. Utopians, urban planners and urban theorists have been testing and using the urban environment as a laboratory for new ways of living and organising the city; Fig. 1. Mario Chiattone in Modern Metropolis 1914, shows a utopian city that is vertical with technological challenges. In 1964, Archigram's Plug-in-Play city is a dream of the mutant city, always growing and never equal to itself, adopting to changes and stimulating creativity. It has also been noted that there are certain references in the ideal city of Campanella, 1602, which is a super positioning of six spheres, like that of the 21st century Smart city today where space dematerialises in to and as networks. Similarly, the Garden City ideology by Ebenezer Howard also constructs a radial city with six spheres. In this manner the Smart City could be a model, a system or ideology that deals with solutions, improvement and evolution of the contemporary city, [9].

The role of innovation has always brought about profound change, for example the Industrial Revolution, the advent of electricity, the invention of the automobile and use of reinforced concrete within construction, [9]. It is at this point that mentioning Le Corbusier and his proposal of the Ville Radieuse is most apt. The city is proposed as an architectural ideology that is closely aligned to the financial and economic conditions of

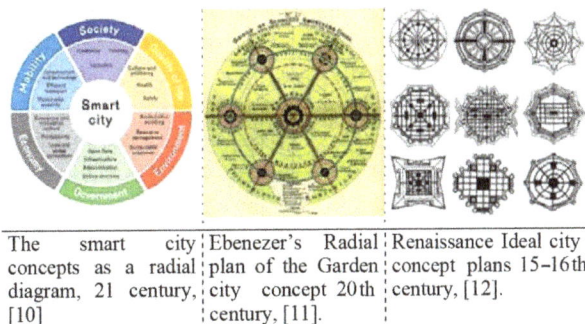

| The smart city concepts as a radial diagram, 21 century, [10] | Ebenezer's Radial plan of the Garden city concept 20th century, [11]. | Renaissance Ideal city concept plans 15–16th century, [12]. |

Figure 1: The concept of Ideal Cities from 15th century to 21st century.

a capital and designed with wide boulevards accommodating for multi lane traffic in both directions with a high density for the residential blocks. Though his radical technical ideas are said to be tamely adapted in to the Ville Radieuse, the city is a proposal to solve the housing crisis within quickly urbanising cities of the 21st century, [2]. Alongside utopias where the latest technological advances are accommodated for, in Le Corbuiser's case, that being the automobile, there are urban theorists who believe strongly in a city designed for people. The likes of Jane Jacobs and Jan Gehl, who advocate a city that is constructed for people as per their needs and a city which in turn shapes the people. In addition, Gehl deals with the same issues as Le Corbuiser of a growing population and economic issues but also environmental issues with focus on sustainability and citizens health. Gehl speaks of the human scale within cities, the 'disgraceful' city spaces all people face where traffic, congestion and pollution have become a norm, cars and carparks have taken over a space that might otherwise be a public park for people to meet and greet within. Instead of using the car, it would be sustainable for citizens health with bicycle lanes introduced, [13].

The smart city in this manner seems like a technological ideology predominantly with many of the aspects that many of the urban theorists and architects have been considering and dealing with for a long time. In this paper, the emphasis is on smart cities that are about people, focused on the inhabitants for their benefit, a city for people, by the people. India being one of the largest democracies in the world with 1.36 billion as a population [14], this paper aims to investigate bottom up approaches, as a method to empower and bring change and benefit to the individuals of a large, vast and diverse population by valuing what is already there within the context of each city. Instead of proposing ideas based on predictions of how inhabitants might or should live, to understand the existing values and shortfalls of a context and to design/ propose as per those requirements, [15].

3.2 Smart Homes in India

The Indian family home has long been known for the joint family ensemble. However, this is fast changing with urbanisation playing a key role in the way family life is evolving. The service class in urban India in the last decade sees a rise in nuclear living. Nuclear living in India is more to do with not living together due to working or studying in another city, it can be described as non-joint living, [16]. An increase in life expectancy has also meant that the elderly population will grow from 91.6 million in 2010, to 157.8 million by 2025. This means that the elderly, in quickly urbanising areas of the country, are vulnerable to living alone and needing care on different levels. It is therefore stated that as the elderly population increases and urbanisation continues life style changes and social challenges faced by the elderly need to be investigated to improve the quality of life that can be led independently, [17].

One such socio-cultural issue within the aging population in India is to continue living at home rather than go to a care home, which is considered shameful. Culturally members of the joint family and even now smaller nuclear families are expected to look after terminally ill elderly family members who might be living away from them. This creates an issue as lifestyles become busier, thereby dependant elderly family members are isolated and neglected in their own homes leading to a low quality of life. So how can homes be developed to allow the elderly person to continue living a life of dignity, and self-care for as long as possible? One ideology that can be useful to consider is that of Le Corbusier; 'the house is a machine for living'. Whilst the idea of architecture as a machine has been criticized for standardising human needs within different contextual issues, standardisation was used as a method to create human wellbeing, [18]. Corbusier states his famous quote in the sense that, 'baths, sun hot-water, cold water, warmth at will, conservation of food, hygiene… an arm chair is a tool for sitting on…', so in short the house is considered to be an efficient tool to help support the delivery of the necessities of life, [19]. Within the smart city paradigm, the 'smart home' can be considered similar in its essence to Corbusier's ideology.

A 'smart home' is a term used for the advanced automated systems that supports inhabitant in daily activities to improve the quality-of-life. A smart home (SH) system is an environment equipped with emerging Internet-of-Things (IoT) technologies. IoT technologies are essentially sensing devices and actuators that are attached to everyday objects in a given environment/space. IoT devices are interconnected and accessible over the internet. Therefore, creating capabilities such as monitoring activities of daily living (ADL), automating and adapting to inhabitant's context to provide just-in-time assistance.

Recent literature has investigated low cost SH solutions such as microcontroller-based that are flexible and scalable [20], and remote access to SH devices [21]. Research in [22] has proposed monitoring and controlling SH environments. It proposes the microcontroller and MQTT cloud platform-based approach for SH monitoring; control (appliances/lighting), detect (intrusion/smoke/gas) and alerting (danger/anomalies). Although, microcontroller-based solutions are flexible, reduce cost and have a higher scalability, there are number of challenges with this approach. One of the key limitations of this approach is that it requires expert knowledge to setup the system and when adding new sensors. The setup process involves three main steps; wiring sensors to microcontroller, programming microcontroller and software system collecting data. Hence, to add new sensors, the three-step setup process need to be repeated. There have been some efforts being made to ease the three-steps setup process for microcontroller-based solutions such as 'over-the-air' programming/firmware upgrade. However, it remains a challenge to create 'plug-and-play' solutions.

Commercial SH kits are now emerging in India and around the world with proprietary and open source components. These kits contain a variety of devices for vision and ambient sensing (i.e. temperature, lighting, switches, motion, and door/window) technologies. For instance, Samsung SmartThings, Insteon and Oakter kit are some of the popular kits available over Amazon India. Other more specialist security kit such as Arlo, iSmartAlaram and SimpliSafe are also available with advance features such as intrusion detection and monitoring. For instance, smart doorbell and smart monitoring cameras that combine camera, motion and sound detection for smart alerting and notification technologies. Most of the kits such as SmartThings, come with a smart hub that support wired and wireless sensors with multiple communication protocol such as ZigBee, Z-Wave, and WiFi. These hubs enable 'plug-and-play' features to add new wireless sensors from different manufacturers effortlessly. Despite the ease of configuring sensors in desired location, the wireless sensors have limited battery life and require frequent replacement. Other devices connected to main power lines are also available that can control lighting and electrical appliances such as Maxico and WeMo switches and plugs. With a vast diversity in sensing technologies and manufacturers, the complexity of interacting with all the devices with individual mobile applications is one key technical challenge. New waves of technology are now emerging such as Amazon Alexa and Google Home that can interact with smart sensors within the SH environment with voice-commands. These speech-based Human–computer Interaction (HCI) capabilities are advantageous for non-technical expert users. However, in the context of India, English is not the first language and detecting popular spoken languages and dialects such as Hindi, Urdu, Bengali,

Marathi and Gujarati remain a research challenge for voice recognition and speech-based technology.

3.3 Ambient Assisted Living (AAL) Systems

Smart home technologies are being utilised to create AAL around the world to support the aging population and people with disabilities [23]. India is advancing to invest in AAL systems with fast urbanisation, nuclear living, and elderly population living alone. AAL encourages independent living by supporting in carrying out Activities of Daily Living (ADL) in their own home. Human Activity Recognition (HAR) is a key part of AAL systems to allow accurate and timely assistance to the inhabitant or carers. The detection of anomalies (i.e. fall), recognising ADLs and changes in daily routines are some of the well-studied research areas in the realm of AAL systems. However, a number of challenges such as multi-occupancy AR, hybrid activity learning and ubiquitous SH monitoring approaches are being investigated by research communities [24]. HAR approaches pave the way for applications in other domains such as healthcare, security, surveillance, smart cities, smart grids, and e-commerce [25].

4 PRIMARY CASE STUDY

4.1 Diu: A Smart City

Diu Island located on the south west coast, off Gujarat, India, is a former Portuguese colony under the Indian government since 1961. Beyond the town walls, lie several villages, Fudam being one, [26]; see Fig. 2. It was reported that nearly 85% of the native local population has migrated to Europe, [27], however many elderly have stayed back in order to continue living in ancestral properties and land.

The elderly lives in their own homes, alone, and are visited by family on a yearly basis. This situation has created many issues from robbery and home safety, to untreated and the onset of chronic illnesses. Recently, CCTV and cameras are being installed by some migrant family members to see their properties and elderly parents in real-time.

4.2 The Migrants Home Village – Fudam

Fudam village is the ancestral village of one of the authors of this paper. It like many other villages on Diu Island is home to the natives who have migrated to Europe. During a series of research trips to Fudam the researcher noticed a serious lack of infrastructure on the island and within individual homes to support the needs of elderly residents like her grandmother. There is an issue with safety within the home and vulnerability. Elderly inhabitants are hard of hearing but do not wear hearing aids due to socio-cultural embarrassment. Grandma's home is relatively small, a one story detached structure set on a large piece of land that allows for both a front and back garden. The toilet is set at the back of the garden. A fading and weakening memory along with hearing issues means the home environment can become dangerous. Additionally, issues such as; guidance over a hob that is left on by mistake, care and assistance after a fall in the home or when unwell are prevalent. Risks of dangerous animals entering the homes, like snakes and monitor lizards are also high due to gardens and neighbouring fields.

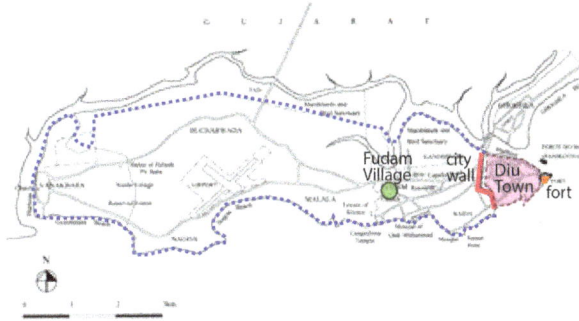

Figure 2: Map of Diu Island showing Fort, Town Wall and Fudam, [26].

4.3 Analysis of Diu for Smart Home System Proposal

Table 1: Table showing opportunities, restrictions and requirements for the SH system.

Social-cultural – *Opportunities (O)*; SH systems can improve the socio-cultural dynamics, giving elderly a better quality of life as they can feel more in control of security of their homes. ***Restriction (R)***; Awareness off and demand for the technology is not there as the elderly don't usually know the latest technology that can help them. The latest technology is not always received in the positive manner as it can be alien and hard to relate to such technologies but also use. ***Requirement (Rq)***; visual elements of the system must be recognised, so the users can get accustomed to the technology quickly or not even realise new technological advances have been integrated in to older devices.

Political – *O*; Smart cities can enable elderly people to live a quality life within their own. ***R***; Budget for private home owners is not available from the council. ***Rq***; No initial financial help, the system needs to be cost effective and conceptually still in this paper but initiate the idea of giving more power to the individual rather than corporates.

Technology – *O*; A different context within this case study allow for innovative solutions to the existing technology in smart home systems. ***R***; Electricity cuts due to maintenance. The technology has to be user friendly and relatable, rather than seen as a new alien technology that the users feels they cannot use. ***Rq***; batteries within sensors will need to be replaced 6-12 months, the system needs to have minimal maintenance.

Environmental – *O*; The tropical climate and the island's geographical location allows for the use of wind or solar power. ***R***; Temperatures and humidity levels are high during different parts of the year and so need to be taken in to consideration for sensors and other parts of the smart home system. ***Rq***; the system should try to utilise wind and solar power, in addition wild animals need to be recognised within the system which can be a problem with a large area of outdoor space to cover.

Economical/Financial – *O*; Many private companies are already investing in SH devices in India. ***R***; Large cost for elderly to afford though migrant family members could afford the SH systems. **Rq**; Cost effectiveness.

Architectural – *O/ R*; The houses are simple in both layout and interior furnishings which allows for the technology to be set up easily. ***Rq***; The smart home system needs to be as non-intrusive as possible so that the lived experience of their home is conserved for the elderly person.

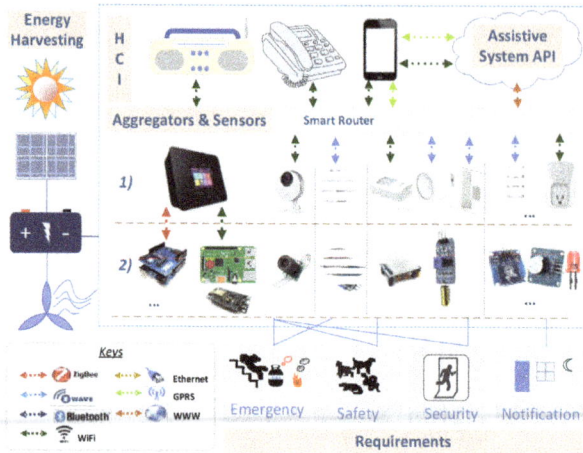

Figure 3: A hybrid edge and cloud computing system architecture of the assistive living system in relation to the socio-cultural primary case study requirements.

5 SOCIO-CULTURAL BASED AMBIENT ASSISTED LIVING

A conceptual view of the proposed AAL and SH technologies approach is depicted in Fig. 3. The approach takes requirements identified in section 4.3 to develop a hybrid computing architecture that is resilient to the power cuts, work with low internet speed and data allowance, cost-effective, and easy to use. To ensure critical components of the SH environment are functional in the event of power cuts, solar and wind energies will be utilised and stored in a battery. An energy monitoring system based on the microcontroller will monitor energy status from the main power grid and switch to battery power energy automatically, [28]. The assistance requirements from the case study can be categorised as emergency, safety, security and notification services. For this, activity recognition with suitable sensing and notification methods for the elderly are highlighted in the following sections.

5.1 Activity Recognition and Pattern Learning

A hybrid activity recognition and pattern learning [29] algorithms are required to adapt to the resident's lifestyle. Although, the proposal for new algorithms are out of the scope of the paper, this paper proposes to delegate these tasks of activity recognition and learning by leveraging the cloud computing and edge computing paradigm. Cloud computing is a centralised approach, which offloads all sensor data collected from the smart home/city environment and the analytics tasks to remote servers. The cloud computing approach is typically used by current AAL systems to store large volumes of sensor data and perform hardware intensive tasks more efficiently. However, the key limitations of cloud computing are that it requires high internet bandwidth to communicate a large quantity of sensor data to the server, which creates a delay in response time, and requires expensive server resources. Therefore, the edge computing paradigm is introduced to create a decentralised approach to delegate some of the computational tasks to aggregators or devices closer to the sensors. Currently, off-the-shelf sensing platforms mainly adapt the cloud computing approach. Hence, the use of bespoke sensing devices will enable the edge computing setup to reduce internet data consumption, increase reliability and continue working even in the event of power outages in the village with the help of the dynamic energy management system.

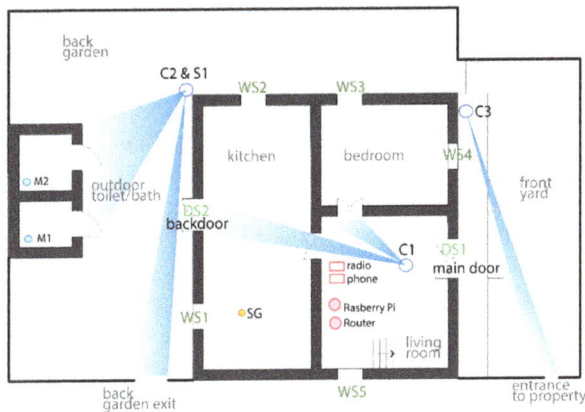

Figure 4: Distribution of smart home environment with sensors and actuators for ambient assisted living system.

5.2 Intelligent Smart Home Environment

The smart home environment will contain a combination of vision and ambient sensing approaches as illustrated in Fig. 4. The main purpose of the vision-based sensing approach is to detect animals, falls, and intruders. Three cameras (C_m) are strategically placed by the door (DS_n) and windows (WS_o) without compromising privacy of the resident as illustrated by the ground floor plan in Fig. 4. C_1 is positioned inside the house to allow viewing of both main doors. C_2 and C_3 allow the coverage of the entrances in to the gardens. C_2 has a speaker (S_1), which allows the doorbell to be heard whilst in the garden. The first floor has another four windows and two doors which would be deployed in the same manner. A low-cost bespoke raspberry Pi and camera-based video processing [30] approach is proposed to detect animals and falls. The live video from a standard camera will allow the raspberry Pi to detect objects and alert the resident. Off-the-shelf cameras are also available at a higher cost and come with a number of advanced features typically used for security and intrusion detection, i.e. motion and sound based recording/alerting, and night vision.

An ambient sensing method will enable personal space to be monitored unobtrusively. For instance, magnetic sensors can detect if the door and windows are open/close and passive infrared (PIR) sensors to detect movement in a given room or garden toilet/bathroom. A low-cost solution will be to deploy with miniature microcontrollers that are wireless and consume low-energy such as ESP8266 12E or Witty Cloud with different types of ambient sensors such as reed switch, smoke/gas (SG_p), temperature (T_q), motion (M_r), sound (S_s), and touch (TO_t). A collection of slave microcontrollers integrated in the telephone and radio are interconnected with the main hub located in the living room to process the sensor data and respond to events automatically. The hub can also selectively publish important data to the cloud web service for behaviour pattern detection overtime and other complex analytics on powerful servers. For this, an assistive system web service will be made accessible over the cloud with application programming interface (API).

5.3 Human-computer Interaction

HCI is a critical factor to consider for engaging elderly with age related problems, limited literacy and no technical skills to use the SH systems being designed. New voice-based speakers such as Amazon Alexa and Google Home technology are becoming increasing popular to control the smart environment and notify residents. However, these smart speakers mainly function on strong Wi-Fi connectivity and do not support native languages. Although, Google Home only recently launched the support for Hindi language for Google Home, more efforts are still needed. These emerging smart devices (i.e. speaker designs, phones, TVs, watches) are not only challenging to operate for elderly with fading memory but they cannot relate to these objects as a friendly tool that they were so used to in past. Therefore, objects such as radio and telephone designs integrated with advance features are proposed to be more relatable objects in the elderly's home environment. An elderly with hearing problems may also feel anxious of being prompted unpredictably without warning, hence, a devotional sound is proposed to be used to get their attention first and then provide any information needed. Therefore, the radio will have storage space for devotional or favourite songs for leisure and notification purposes, and wireless connectivity with the main hub. The landline telephone will be optimised for speed dialling a set of contacts and emergency services and have additional buttons with LEDs flashing to dial a pre-defined contact number. For this, both radio and telephone will be customised using the raspberry pi.

6 DISCUSSION & EVALUATION

India's 100 smart cities mission is a long way to serving private homeowners directly to improve quality of life and independent living in their own home instead of going to care homes at old age. There are little to no schemes available from the government or local councils for the elderly to invest in SH systems. The overall costs of installing AAL systems with advanced smart home devices are still high and unfeasible for the elderly in some parts of India. In general, the benefits of these technological tools such as improved safety, security and support in ADLs are still not well unknown or conveyed to the elderly. Consequently, greater awareness of such technology will increase their ability to embrace them.

New waves of assistive technology are still in their infancy and companies must tailor their products so that they are relatable and enhance existing objects in the home environment to specific stakeholders such as the elderly. Therefore, this paper revisits old relatable devices such as radio and wired telephone for the elderly to feel less alienated with the new technology and age-related challenges such as weakening cognitive ability and fading memory. There are several technical and practical challenges in developing a mature, one AAL solution. For instance, developing sensing devices that can be self-discoverable, self-powering [31] and unobtrusively integrated into our environment or worn body with the likes of sound [32], and smart clothing/textile [33] technologies.

7 CONCLUSION AND FUTURE WORK

In this paper, it can be understood that using existing technologies within the homes of the elderly along with customising and connecting familiar technologies from the elderly person's experience a smart home system can be designed around specific socio-cultural and environmentally inherent issues. Many issues exist within the proposal made in this paper. The social, cultural and wider critical thoughts regarding the conceptual proposal in this paper are; (1) currently, grandma in the case study is used to having two cameras in the house already, one in the living room and one in the garden. Cameras are considered intrusive and unethical at times, depending on where they are installed.

The proposal in this paper uses existing cameras and grandma's familiarity with them. But in other contextual situations, cameras might not be utilised as liberally as done in this case. (2) In different contexts, different current technologies like smart phones and smart TV can be more appropriate. Currently, grandma has a standard TV, which she only watches an hour a day and a smart phone which she doesn't know how to use. However, the use of smart phones and smart TV's by the other elderly in other regions might be higher and can be integrated in the SH system. (3) Also, in some parts of India, villages are still struggling with electricity, water and home sanitation which becomes an important factor in the overall quality of life for the elderly, which could pose more interesting problem contexts to deal with.

Methodologically, future research should take a larger sample of case studies/scenarios within a particular city, or urban environment to understand the diversity of the needs of the elderly living within the city/area. Moreover, a prototype will be developed in future work, based on the proposed approaches.

REFERENCES

[1] Ismail, N., *Smart Cities in India: Embracing the Opportunity of Urbanisation*, 2018.

[2] Teige, K. & Dluhosch, E., The minimum dwelling [Internet]. Chicago; 2002. (Mit Press). available at https://books.google.co.uk/books?id=3Q4TYPRPKpUC

[3] Hinkelmann, K. & Witschel, H.F., *How to Choose a Research Methodology*, Univeristy Appl Sci Northwest Switzerland, Sch Bus. 2009.

[4] Meggi, A., Representing the colony: Documenting the other perspective. In *Historical Perspectives Global Communities Conference*,. Kelvin Hall, Glasgow, 8–9 June 2018.

[5] Meggi, A., Towards a digital heritage: Evaluating methods of heritage interpretation, diu town—a case study. *International Journal of Heritage Architecture: Studies, Repairs and Maintence [Internet]*, **2(3)**, pp. 406–416, 2017. available from http://www. witpress.com/doi/journals/HA-V2-N3-406-416

[6] Grossi, G. & Pianezzi, D., Smart cities: Utopia or neoliberal ideology? *Cities*, **69**, pp. 79–85, 2017. https://doi.org/10.1016/j.cities.2017.07.012

[7] International P. Smart cities: Utopian Vision, Dystopian Reality. 2017;(October).

[8] Anto, A. & Dhwani P., How the unfinished city of Lavasa became a nightmare for Indian banks | Business Standard News [Internet]. available at https://www.business-standard. com/article/current-affairs/how-the-unfinished-city-of-lavasa-became-a-nightmare-for-indian-banks-118061900095_1.html, 2018 (accessed 7 February 2019).

[9] Mata, A.M., *Is Smart City an Utopia ? Lessons Learned and Final Reflections*, July 2018.

[10] Brussels Smart city [Internet]. available at https://smartcity.brussels/the-project-definition 2019 (accessed 7 Febrauary 2019).

[11] Grand Reductions: 10 Diagrams that Changed City Planning [Internet]. Urbanist Article. available at https://www.spur.org/publications/urbanist-article/2012-11-09/ grand-reductions-10-diagrams-changed-city-planning, 2012.

[12] Grand Reductions: 10 Diagrams that Changed City Planning. Urbanist Article. 2012.

[13] Gehl, J. & Rogers, R., *Cities for People [Internet]*. Island Press, available at https:// books.google.co.uk/books?id=lBNJoNILqQcC, 2013.

[14] India Population [Internet]. available at http://www.indiapopulation2019.in/ 2019 (accessed 7 February 2019)

[15] Lee, T. & Jane, J., Bottom-Up Thinker [Internet]. available at http://timothyblee. com/2010/07/13/jane-jacobs-bottom-up-thinker/, 2013 (accessed 7 Febrarury 2019).

[16] Patel, T., The Family in India: Structure and Practice [Internet]. Sage Publications; (Themes in Indian sociology). available at: https://books.google.co.uk/books?id=K-l_Ve_GKOIC, 2005.

[17]   B Mane A. Ageing in India: Some Social Challenges to Elderly Care. *J Gerontol Geri-atr Res [Internet]*; available at http://www.omicsgroup.org/journals/ageing-in-india-some-social-challenges-to-elderly-care-2167-7182-1000e136.php?aid=69369, 2016.

[18]   Atmodiwirjo, P. & Yatmo, Y.A., Architecture as machine: Towards an architectural system for human well-being. *LC2015—Le Corbusier, 50 Years Later*, pp. 1–10, 2015.

[19]   Morse, G., What is a Machine for living in? [Internet]. available at https://placeex-ploration.com/2015/10/28/a-house-is-a-machine-for-living-in/, 2015 (accessed 7 February 2019).

[20]   Nath, P. & Pati, U.C., Low-cost android app based voice operated room automation system. *In 2018 3rd International Conference for Convergence in Technology (I2CT)*, IEEE, pp. 1–4, 2018..

[21]   Shailendra, E. & Bhatia, P.K., Analyzing Home Automation and Networking Technolo-gies. *IEEE Potentials [Internet]*, **37(1)**, pp. 27–33, 2018. available at http://ieeexplore.ieee.org/document/8253757/

[22]   Malche, T. & Maheshwary, P., Internet of Things (IoT) for building Smart Home System. *In IoT in Social, Mobile, Analytics and Cloud*, pp. 65–70, 2017.

[23]   Banks, B.J., The 'age' of opportunity. *IEEE Pulse*, **8(2)**, pp. 12–5, 2017.

[24]   Ihianle, I.K, Naeem, U., Syed, I. & Tawil, A.R., A hybrid approach to recognising activities of daily living from object use in the home environment. *Informatics [Inter-net]*, **5(1)**, pp. 1–25, 2018, available at http://www.mdpi.com/2227-9709/5/1/6

[25]   Aouedi, O., Anis, M., Tobji, B. & Abraham, A., Internet of things and ambient intelligence for mobile health monitoring : A review of a decade of research. *Int J Com-put Inf Syst Ind Manag Appl.*, **10**, pp. 261–270, 2018.

[26]   Shokoohy, M. & Shokoohy, N.H., The island of diu, its architecture and historic remains. *South Asian Stud [Internet]*, **26(2),** pp. 161–191, 2010. available from: http://dx.doi.org/10.1080/02666030.2010.514743

[27]   Rahman, A.P., With the island's old-timers moving to Europe, Diu's 400-year-old Portuguese influence is fading [Internet]. *The Hindu*. available at https://www.thehindu.com/society/with-the-islands-old-timers-moving-to-europe-dius-400-year-old-portu-guese-influence-is-fading/article23528249.ece, 2018 (cited 2 February 2019).

[28]   Srivastava, P., Bajaj, M. & Rana, A.S., IOT based controlling of hybrid energy system using ESP8266. 2018 IEEMA Eng Infin Conf eTechNxT 2018, pp. 1–5, 2018.

[29]   Chen, L., Nugent, C. & Okeyo, G., An ontology-based hybrid approach to activity modeling for smart homes. *IEEE Trans Human-Machine Syst*, **44(1)**, pp. 92–105, 2014.

[30]   De Miguel, K., Brunete, A., Hernando, M. & Gambao, E., Home camera-based fall detection system for the elderly. *Sensors (Switzerland)*, **17(12)**, 2017.

[31]   Lin, Z., Yang, J., Li, X., Wu, Y., Wei, W., Liu, J., Chen, J., & Yang, J., Large-scale and washable smart textiles based on triboelectric nanogenerator arrays for self-powered sleeping monitoring. *Adv Funct Mater [Internet]*, **28(1)**, p. 1704112, 2018. available at https://onlinelibrary.wiley.com/doi/abs/10.1002/adfm.201704112

[32]   Giannakopoulos, T. & Konstantopoulos, S., Daily activity recognition based on meta-classification of low-level audio events. *ICT4AWE 2017 – Proc 3rd Int Conf Inf Commun Technol Ageing Well e-Health*. 2017.

[33]   Wen, Z., Yeh, M.H., Guo, H., Wang, J., Zi, Y., Xu, W., Deng, J., Zhu, L., Wang, X., Hu, C., Zhu, L., Sun, X., & Wang, Z. L., Self-powered textile for wearable electronics by hybridizing fiber-shaped nanogenerators, solar cells, and supercapacitors. *Science Advances [Internet]*, **2(10)**, 2016, available at http://advances.sciencemag.org/con-tent/2/10/e1600097

STRATEGIES FOR THE DEVELOPMENT OF THE VALUE OF THE MINING-INDUSTRIAL HERITAGE OF THE ZARUMA-PORTOVELO, ECUADOR, IN THE CONTEXT OF A GEOPARK PROJECT

GRICELDA HERRERA FRANCO[1], PAÚL CARRION MERO[2], FERNANDO MORANTE CARBALLO[2], GEANELLA HERRERA NARVÁEZ[3], JOSUÉ BRIONES BITAR[3] & ROBERTO BLANCO TORRENS[4]
[1] Universidad Estatal Península de Santa Elena, Facultad de Ciencias de la Ingeniería, La Libertad, Ecuador
[2] ESPOL Polytechnic University, Escuela Superior Politécnica del Litoral, Centro de Investigación y Proyectos Aplicados a las Ciencias de la Tierra, Guayaquil, Ecuador
[3] BIRA Bienes Raíces S.A., Zaruma, Ecuador
[4] Instituto Superior Minero Metalúrgico, ISMM, Facultad de Geología y Minas, Cuba

ABSTRACT
The enhancement of heritage resources helps to promote conservation, contributes to more significant and better protection, and favors the efficient use of these resources. Many heritage works and liabilities linked to mining activity are abandoned, causing the deterioration of heritage resources that may become environmental liabilities over time. This work aims to develop strategies for the development of the Mining-Industrial heritages through participatory methods for geomining enhancement and development of places in the Zaruma-Portovelo area. The proposed methodology consists of: (i) The creation and development of a database with several publications and documents that register the Mining-Industrial heritage sites; (ii) The assessment of mining-industrial sites based on criteria or methodologies proposed by other authors; (iii) Focal group work considering: (a) The identification and cataloging geosites of interest, (b) The creation and/or development of museums, tourist mines, mineral routes or geoparks in which the natural and geological-mining factors complement each other and (iv) SWOT analysis and matrix which provides several strategies for value-making of geomining heritage and its promotion in the development of geotourism in a project proposal for Zaruma and its surroundings. In conclusion, this work includes twelve unified mining sites in the Proposal for Geopark 'Ruta del Oro,' as a strategy to guarantee the conservation of heritage values and contribute to local development and geotourism.
Keywords: conservation, environmental liabilities, geosite, geotourism, Mining-Industrial heritage.

1 INTRODUCTION
Heritage constitutes the historical memory of a nation and it is the fundamental pillar to build a culture that stands out in attitudes, behavior and implicit values acquired by a man, as a member of a society over time. These values are the result of their relationship with the environment, which makes it valid for sustainable development and reveal through assets with an attractive appeal, either because of its outstanding artistic value or its originality, rarity or extravagance [1]. In this area of study, the population plays an essential role in the contribution of the image given to a site; therefore, rescuing the heritage derives from a committed perspective that freezes 'valuable' situations. Restorations 'value' the elements considered the most attractive for highlighting the authenticity of a place. The knowledge of history has all the elements of the awareness process that a community has [2]. Validating that knowledge is a complex task because it comprises conservation strategies and heritage interpretation. These elements consider the value-added defined as the interpretation and presentation of different values given to the patrimony such as material, symbolic, emotional, social, and educational or use value [3].

© 2020 WIT Press, www.witpress.com
DOI: 10.2495/EQ-V5-N1-48-59

The existence of a culture without heritage and a society without memory is impossible [4]; that is why, the cultural heritage of a country or region is constituted by all those elements and tangible or intangible manifestations produced by societies, result of a historical process that identifies and distinguishes that country [5, 6]. However, social value is the key that makes heritage an essential reality for the understanding of culture through societies capable of transmitting their value to future generations. There are different activities for which the heritage is reflected. In this case, we will focus on the industrial activities that have provided over the years a great importance in different mining areas, which have led the creation of an industrial heritage closely related to the mining heritage. It includes surface and subsurface level, hydraulic installations, transport, machinery, documents or objects related to previous mining activities with historical, cultural or social value [5, 7, 8]. This s value allows access to new geotouristic destinations for the sustainable development of a region.

In Ecuador, there are many places with outstanding features to mining sites. In El Oro province, precisely the district Zaruma-Portovelo is a study area, where mining activity stands out since the Pre-Inca stage. In the fifteenth century, the Inca Tupác Yupanqui was sent to the conquest of the north of his empire [9, 10]. In the time of the colonization, the Spaniards ground the quartz for obtaining the mineral and extracted gold of three pounds of weight given as a gift to the King of Spain. The provenance was from the 'El Sexmo' gold mine [8]. The rehabilitation of this mining area began in 2002 by the company BIRA S.A. After this initiative, 'El Sexmo' may become a cultural tourist attraction that associates mining with a producing industry for the country development [11]. Recently, there has been a resurgence of mining in megaprojects that have prompted the Ecuadorian government interest in these social sectors [12].

This study seeks to rescue Zaruma-Portovelo historical values and geotourism places, where mining activity plays a vital role. Mining phases of prospecting, exploration, benefit, smelting, refining and marketing of minerals [5, 7, 13, 14] are used for local profit-making purposes. Taking advantage of this knowledge, this study advocates labor reintegration after the closure of mines as a patrimony to promote geotourism along with the relocation of human resources in existing mines [15]. Furthermore, the industrial heritage comprises all the material and immaterial vestiges inherited by industrial activities. Industry is the production of an item for its commercialization, not for self-consumption [16]. The industrial-mining heritage becomes the architectural, museum or recreational point of view along with the rehabilitation of areas damaged by the use of post-mining land. This gives value not only to the old industrial establishments but also to the vast industrial facilities and the scientific establishments that become the object of attention and visit as indicated by the Spanish geographer Horacio Capel in [17]. This corresponds to the mining-industrial heritage where activities related to geotourism are addressed to achieve sustainable development.

The interest in rescuing Zaruma and Portovelo mining-industrial heritage is based on the proposal Geopark 'Ruta del Oro'. We considered twelve geosites of interest, which contribute to the enrichment of historical knowledge, education, cultural identity and economic reactivation of the populations [18, 19]. These locations have benefited from the mining activity with the creation of development strategies compatible with their own history and transcendence.

These strategies arise from alternative activities such as the creation and development of museums, tourist mines, mineral or Geopark routes in which the natural and industrial-mining factors complement each other. Mining sustainability is achieved through policies that allow communities to take advantage of this type of activity after depleting the deposits, the knowledge acquired and the infrastructure [20]. The process takes place, through the enhancement of mining interest sites, whose development paths are focused on the analysis of the positive impacts that mining generates on the environment. Another consideration is the compensation of irreversible impacts in order to exploited potential cultural heritage that tourists give to the tangible and intangible heritage of industrialization [21, 22].

2 STUDY AREA

El Oro province, located in the south of Ecuador, is divided into three zones: the coastal plain or lower part, the upper part and the Jambelí archipelago. The second zone is called the Orense Plateau because it is located in Los Andes Mountain Range, characterized by ancient volcanism. These mountain ranges are undulated and it is formed by sites such as Saracay, Balsas, Marcabelí, Piñas, Atahualpa, Portovelo, Zaruma up to the ridges of the Hoya Puyango [18]. The study area is Zaruma and Portovelo, the name Zaruma comes from two Quechua words SARA: head and UMA: corn (head of corn), and it was founded by Alonso de Mercadillo, the same founder of Spain by order of the Spain King [19]. Zaruma and Portovelo are the oldest mining region in the country because they belongs to one of the five mining districts of more considerable antiquity and importance. The district is limited to the North with Guayas and Azuay provinces, South and East with Loja and Northwest with Guayaquil Gulf [20, 24].

Currently, Zaruma has 659.40 km^2 of extension and a population of 24097 inhabitants [18]. Its auriferous wealth is the primary source of work for inhabitants and immigrants. In Zaruma, 24.11% of the population works in mines equivalent to 2369 jobs. In Portovelo, 32.85% are dedicated to mining with 1692 jobs, which make a total of 4061 jobs [21]. Because in the Zaruma sector there are non-technical mining practices and outside the law which have caused the mines that were in operation have been closed, since they have caused environmental and social impacts. Therefore, the Ecuadorian government has decreed an extension of the mining exclusion zone. This fact emphasizes that in the Zaruma sector there has been no technical mine closure process, and this process, in the current situations, becomes essential. The closure of mines is a technical process that from its inception foresees the conditions in which the authorized sectors must be enabled for later use and in the best safety conditions, so that the closure of mines is closely linked to the processes of clean energy production.

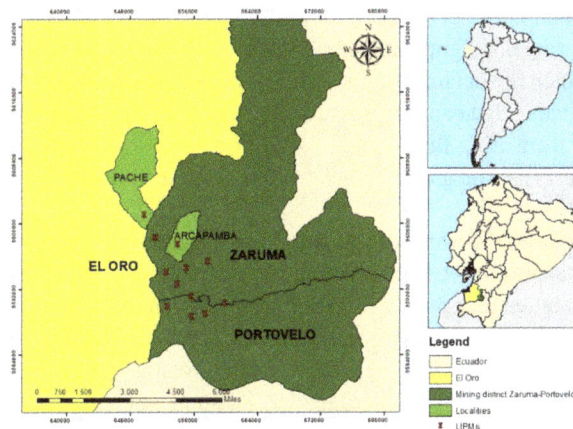

Figure 1: Zaruma-Portovelo mining-industrial sites [23].

Therefore, we have seen the need to look for other sources of income such as the commercial, agricultural, livestock, craft, and tourism sectors, being tourism our subject of interest. For this reason, in this work are considered twelve mining-industrial sites rich in heritage that promote geotourism in the region, such as Mina Turística el Sexmo, Museo Geocientífico y Mineralógico Magner Turner, Museo Municipal de Zaruma, Escultura en Honor al minero, Monumento al minero en Portovelo, Manantiales de aguas termales Portovelo, Mina Vizcaya, Minas Antiguas de Miranda, Minas de Oro Reina de Fátima, Antigua Planta de Beneficio de la SADCO, Plantas de beneficio Vía Portovelo-Pache and Planta Hidroeléctrica El Pache. 'Figure 1 shows the mining-industrial sites of the study area'.

3 METHODOLOGY

To ensure the results of this study, a methodology composed of four phases was proposed: (i) The creation and development of a database, publications and others that register mining-industrial heritage sites, thus ensuring up-to-date information from various authors; (ii) The assessment of mining-industrial sites was based on methodologies proposed by authors such as Esther et al., [22] and the methodology of Inventario Español de Lugares de Interés Geológico (IELIG) [5]. The first corresponds to a methodological guide for the integration of mining heritage in the environmental impact assessment, where Mining Heritage Sites called LIPMs, which include the industrial part, covers two factors: (1) LIPM intrinsic value, making relevance to the mining-industrial aspects and (2) use-value of the LIPM. In this way, with the collaboration of experts such as Magner Turner, Tito Castillo, Vicente Figueroa and Carmen Macas, together with the necessary fieldwork, we obtained the assessment of the sites global interest (Vg) or merit of conservation through the establishment of weights given to each one according to [22]. The elements that are equal or greater than 300 points will be considered 'Very high' mineral heritage places of interest. With values between 300 and 200 will be of 'High' interest. The scores between 100 and 200 will be of 'Medium' interest, and the elements that do not reach the 100 points will be of 'Low' interest.

Subsequently, the change in the value of a LIPM is estimated after the execution of a project, resulting in the loss of its conservation value in the following categories: remarkable, significant, moderate and slight value, being 'remarkable' the consequence of a substantial improvement of its intrinsic value or its possibilities of use. According to these criteria, the resulting state of impact is obtained based on the two previous results together with the Regulation of Environmental Impact Assessment (EIA). The state obtained for each of the LIPM will be qualified as: state not significant, compatible, moderate, severe and critical, where the latter is the one in which the decrease in the value of the LIPM is comparable with its destruction or with a loss of value that cannot be assumed [22]. Likewise, the methodology developed in [5], states that by measuring the degree of scientific interest (Ic), educational interest (Id) and the tourist interest (It), the global interest value, similar to the previous methodology, is obtained.

The phase (iii) includes the work of a focal group considering: (a) The identification and geosites of interest cataloging, (b) The creation and development of museums, tourist mines, mineral or Geopark routes. For the presentation of results, in phase iv) the preparation of an analysis of Strengths, Weaknesses, Opportunities and Threats (SWOT) and its corresponding matrix TOWS is proposed, in which several strategies for the implementation of the value enhancement of the mining-industrial heritage and its impact on the growth and development of geotourism in a project proposal for Zaruma. Figure 2 summarizes the methodology proposed in this study.

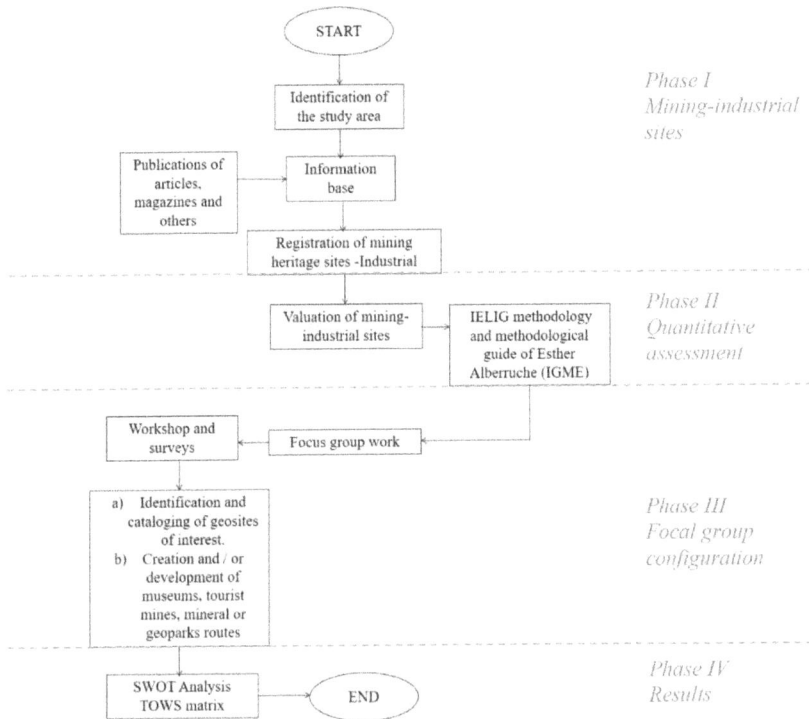

Figure 2: Methodology flowchart used for the research.

4 RESULTS

After identifying the area of this study and having compiled all the bibliographical information available, twelve geosites of mining-industrial interest could be registered and presented in Table 1.

The evaluation of these places was done through surveys of experts from the sites of interest. The researchers applied the methodology proposed in [22] to find the global value (Vg) of each of the geosites to be considered LIPMs at present. They are valued in the category of 'Very high' interest to Plantas de beneficio Vía Portovelo-Pache, Antigua planta de beneficio de la SADCO, Minas antiguas de Miranda, Museo Municipal de Zaruma, Museo geo-científico y mineralógico Magner Turner y la Mina turística 'El Sexmo' (Fig. 4) with a score of 330, 300, 307.5, 320, 330 and 300, respectively. To 'High' interest category belong Planta Hidroeléctrica El Pache, Minas de Oro Reina de Fátima, Mina Vizcaya, Manantiales de aguas termales Portovelo, Monumento al minero en Portovelo and Escultura en honor al minero en Zaruma with a score of 245, 251.5, 277.5, 244, 237.5 and 237.5, respectively. The LIPMs selected are in the two highest categories, which make them conservation icons (Fig. 3).

Once the LIPMs Global Value has been obtained, they go through the calculation of the loss of their conservation value imposed by the execution of a project. In this new evaluation of the LIPM, Antigua Planta de beneficio de la SADCO with 37%; la Planta Hidroeléctrica El Pache, Planta de beneficio Vía Portovelo-Pache, Minas de oro Reina de Fátima, Mina Vizcaya and Manantiales de aguas termales Portovelo with 27.25% entered the 'Significant' category. In the 'Moderate' category, we found Minas Antiguas de Miranda with 24.5% and

Table 1: List of mining-industrial sites of interest and their location in the study area.

N°	Geosite	Main interest	Location	Canton
1	Mina Turística el Sexmo	Mining	Zaruma	Zaruma
2	Museo Geocientífico y Mineralógico Magner Turner	Mining	Portovelo	Portovelo
3	Museo Municipal de Zaruma	Mining	Zaruma	Zaruma
4	Escultura en Honor al minero	Mining	Zaruma	Zaruma
5	Monumento al minero en Portovelo	Mining	Portovelo	Portovelo
6	Manantiales de aguas termales Portovelo	Mining	Portovelo	Portovelo
7	Mina Vizcaya	Mining	Zaruma	Zaruma
8	Minas Antiguas de Miranda	Mining	Zaruma	Zaruma
9	Minas de Oro Reina de Fátima	Mining	Arcapamba	Zaruma
10	Antigua Planta de Beneficio de la SADCO	Industrial	Portovelo	Portovelo
11	Plantas de beneficio Vía Portovelo-Pache	Industrial	Pache	Portovelo
12	Planta Hidroeléctrica El Pache	Industrial	Portovelo	Portovelo

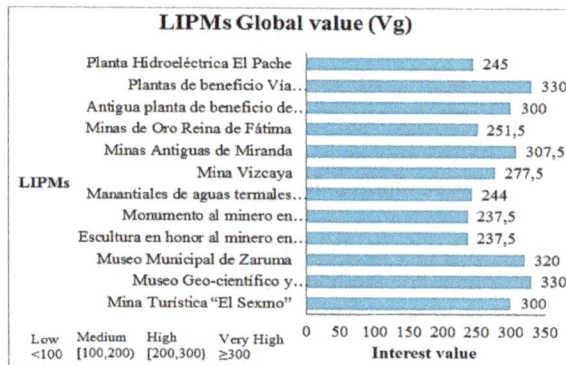

Figure 3: Global value (Vg) results of the mining heritage sites.

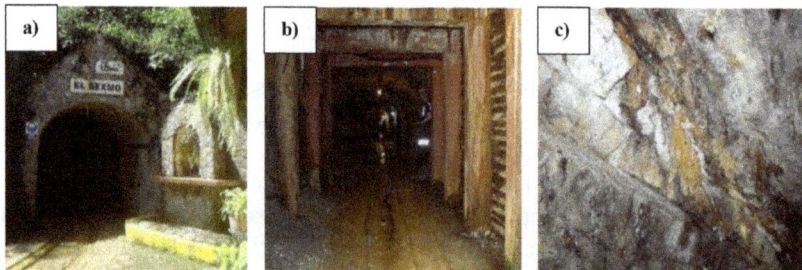

Figure 4: Example of mining site (Tourist Mine 'El Sexmo') registered in the inventory. (a) Entrance to the pit; (b) underground passage; (c) gold veins inside the mine.

the Museo Geo-científico y Mineralógico Magner Turner with 11.25%. Finally, in the 'Slight' category is found Monumento al minero en Portovelo, Escultura en honor al minero en Zaruma and Mina turística 'El Sexmo' with 5.5% (Figs. 5 and 6).

The 'significant' loss of value of LIPMs is the consequence of a sensible modification of the LIPM with appreciable repercussions, in short, or long term, on its intrinsic value or possibilities of use. In those sites with a 'moderate' loss of their value, the effect may occur in the future as repercussions on their intrinsic value or possibilities of use, while those of 'slight' loss of value have slight repercussions in the future [22].

To complement this methodology, we estimated the change in value that the LIPM may take before and after the project execution and the loss of its conservation value. According to the matrix in [22], to estimate this impact, Figs. 3 and 5 are used, categorizing the geosites in the following states described in Table 2. The LIPM tolerance to negative alterations associated with a project or activity depends on the global value (Vg) given to the LIPM. Therefore, the acceptable threshold of loss of value will be lower in the LIPM of very high and more permissive or higher in those of low value, as indicated in Table 2.

The status of each geosite can be critical, severe and moderate. For example, in 'critical' status, we found Antigua Planta de Beneficio de la SADCO due to the total abandonment of the site and the total deterioration of its infrastructure by the action of external agents. Likewise, the Plantas de beneficio Vía Portovelo-Pache are in 'critical' statusdue to the excessive pollution of its rivers, which causes a decrease in the value of the LIPM that is understood as the destruction or total loss of its conservation value, that is, permanent without recovery. In 'severe' status are the Manantiales de aguas termales Portovelo, Mina Vizcaya, Minas Antiguas de Miranda, Minas de oro Reina de Fátima and Planta Hidroeléctrica El Pache, where, the reduction of the LIPM value is evident in infrastructure, and the recovery of the conditions of the LIPM requires the adaptation of significant protective measures. In the 'moderate' status are Mina turística 'El Sexmo,' Museo Geo-científico y Mineralógico Magner Turner, Museo Municipal de Zaruma and Plantas de beneficio Vía Portovelo-Pache.

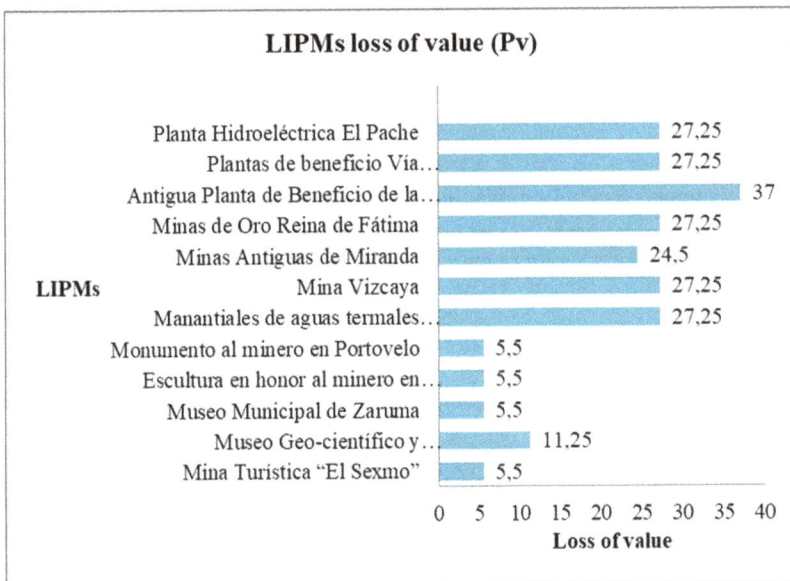

Figure 5: Loss of value results of mining heritage sites (LIPMs).

Figure 6: Industrial site example with significant value loss. (a) Old American Sink of the SADCO; (b, c) El Pache hydroelectric plant.

Table 2: Relationship of the status of the LIPM based on its loss of value due to external effects.

N°	LIPMs	Global Value (Vg)	Loss of Value (Pv)	Status
1	Mina Turística 'El Sexmo'	Very High	Slight	Moderate
2	Museo Geo-científico y Mineralógico Magner Turner	Very High	Moderate	Moderate
3	Museo Municipal de Zaruma	Very High	Slight	Moderate
4	Escultura en honor al minero en Portovelo	High	Slight	Compatible
5	Monumento al minero en Portovelo	High	Slight	Compatible
6	Manantiales de aguas termales Portovelo	High	Significant	Severe
7	Mina Vizcaya	High	Significant	Severe
8	Minas Antiguas de Miranda	Very High	Moderate	Severe
9	Minas de Oro Reina de Fátima	High	Significant	Severe
10	Antigua Planta de Beneficio de la SADCO	Very High	Significant	Critical
11	Plantas de beneficio Vía Portovelo-Pache	Very High	Significant	Critical
12	Planta Hidroeléctrica El Pache	High	Significant	Severe

Forthese sites, it is suggested to establish non-intensive protective or corrective practices. Finally, Escultura en honor al minero and Monumento al minero in Portovelo are in a 'compatible' status, that is, their loss is acceptable.

5 DISCUSSION

The geosites and mining sites value-setting approach is reinforced by the mines closure criterion since the use of these sites as geotourism enters a very important demonstrative process what is the mines closure. Geopark proposal comprises an integrative project that seeks to make use of the previously exploited areas, and thus enabled them for new alternatives, without losing the relationship with the sector mining identity. In this way, the areas considered

Table 3: TOWS analysis of the workshop 'Geotourism prospects in the upper part of El Oro province, Zaruma-Portovelo'.

Strengths	Opportunities
Zaruma is considered the first Ecuador Magical Town. The sites of mining-industrial interest in the area are accessible and with high global heritage value. Compatibility of conservation in the historical and cultural wealth of the country. Mining technical knowledge to be exploited. Portovelo highlights its mining history with the American company SADCO. The mining resources in the area link the community with its history.	Expansion of tourism. Protection and restoration of the geosites to study and disseminate the mining-industrial history of the country. Proposals for base projects for future initiatives of a Geopark ('Ruta del Oro') Through the project Geopark 'Ruta del Oro', work sources are created. Projects of governmental entities and agreements with universities for the recovery of geosites and promulgate geotourism. A tourist development susceptible to improve.
Weaknesses	**Threats**
Despite its geotouristic importance, certain sites look neglected. Little support from government entities for the conservation of these geosites. Weak conservation awareness in places of relevance. Lack of teaching content and links with interested universities for development plans. Lack of informative content in mining-industrial heritage places of interest. Lack of 'social awareness' that requires industrial heritage.	Extinction of sites of mining-industrial heritage interest. Contamination of water sources due to bad mining practices that affect the population. Lack of investment to preserve, restore and value the Heritage. Geosites abandoned and little valued by the community. High level of vulnerability in places of industrial heritage interest and loss of value. Problems related to the use of the land and its natural environment.

in the mine-closing process are enabled and are a major factor in the cleaner energy production as mining infrastructure and galleries can be very important in providing: (a) safe sites for alternative uses, (b) runoff outlets from treated and environmentally friendly waters, (c) uses of spaces for the development of the city and the sector, (d) use of underground spaces for geotourism, (e) use of waterfalls for hydroelectric power production, (f) use of old galleries for ventilation systems, and other alternatives and uses that register an environment in good condition and a clean energy production for the sector.

The results presented through a SWOT analysis were obtained from the focal group work and its feedback with the Sustainable Development Goals (SDG) and geotourism perspectives issues in the upper part of El Oro province, Zaruma-Portovelo.

The result of the analysis in Table 3 is a useful tool for the preparation of the TOWS matrix in Table 4, which summarizes the following general strategies, considering internal characteristics (strengths and weaknesses) and external characteristics (opportunities and threats).

Table 4: TOWS matrix of the Workshop 'Geotourism prospects in the upper part of the El Oro province, Zaruma-Portovelo'.

Strategies: Strengths + Opportunities	Strategies: Weaknesses + Opportunities
2.a,b Restorations that 'put in value' the elements considered of greater attraction to highlight the authenticity of the geosites and expand the tourist offer. 3.e,f Encourage the authority's commitment and responsibility for the formulation of tourism projects with short-term implementation plans. 6.c,d Implementation of new tourism initiatives for the local development of the mining area such as the creation of a geopark.	1.f Adapt the infrastructure of services annexed to the mining site (roads, signage, recreational areas, etc.) to improve the quality of life of the population and the satisfaction of tourists during their stay. 4.b Development of programs by educational institutions, public and private companies in the sector to raise awareness of the values of industrial-mining heritage, its importance of recovery, conservation, and dissemination. 5.a To make agreements with universities for management works such as georuts, mine theme parks and geoparks.
Strategies: Strengths + Threats	**Strategies: Weaknesses + Threats**
4.c Work of multidisciplinary groups to contribute with a plan for the recovery of mining-industrial sites and their enhancement for sustainable development. 5.f Encourage the consolidation of mining-industrial places of interest according to the conservation status validated in this work to enhance them as tourist destinations.	5.a.e Implement promotion strategies (marketing) so that these sites are not forgotten by the community itself and disseminate its informative content and importance in the country's mining history. 2.c Look for investors to develop a proposal that comes from the community itself and reveals their identity.

6 CONCLUSIONS

The study reveals the existence of places of high mining-industrial heritage interest in the Zaruma-Portovelo district. It follows the example of similar initiatives launched in some European countries. These sites could be exploited through local development strategies. In this way, it is essential to take adequate legal and financial measures to materialize the viability of mining-industrial uses in any of the exploitation figures. Thus, the sites could contribute to socio-economic and geotouristic development. In mining, the extraction of gold and the problems derived from related activities such as environmental problems and land destabilization require alternative development strategies such as those mentioned in the 'Discussion' section. Geotourism represents a sustainable activity, which is also compatible with other current socio-economic activities in the area. For this purpose, it is necessary to raise awareness among the population through educational programs, workshops, and talks related to heritage and culture. These strategies could serve as an incentive to promote what is known, they could be the key to sustainable socio-economic and environmental development.

REFERENCES

[1] Campoverde, C., Ramírez, G., Carrión, P. y Herrera, G., Zaruma-Portovelo: Contexto geominero de patrimonio y diversidad para el desarrollo sostenible. Libro de Actas del Cuarto Congreso Internacional sobre geología y minería ambiental para el orde-

namiento del territorio y el desarrollo, pp. 163–186, ed. Geoparque de la Comarca de Molina-Alto Tajo, SIGMADOT Y SEDGYM, 2016. (In Spanish)

[2] Ramírez G., Campoverde C. y Carrión P., Potencial geoturístico-minero de la Ruta del Oro, Ecuador. Libro de Actas del Cuarto Congreso Internacional sobre geología y minería ambiental para el ordenamiento del territorio y el desarrollo, pp. 213–230, ed. Geoparque de la Comarca de Molina-Alto Tajo, SIGMADOT Y SEDGYM, 2016. (In Spanish)

[3] Boivin M. & Tanguay Georges A., Analysis of the determinants of urban tourism attractiveness: The case of Québec City and Bordeaux. Journal of Destination Marketing & Management XI, ScienceDirect, pp. 67–79, ed. Elsevier, marzo, 2019. https://doi.org/10.1016/J.JDMM.2018.11.002

[4] Sanmartín G. M., Zhigue A.R. y Muññoz G., Environmental managemet for the conservation of Zaruma, historical and cultural heritage of Ecuador. *Revista Conrado*, **13** (1-Ext), pp. 116–121, 2017.

[5] Carrión P., Herrera G., Briones J., Caldevilla P., Domínguez M.J. y Berrezueta E., Geotourism and Development Base on Geological and Mining Site Utilization, Zaruma-Portovelo, Ecuador, *Geosciences*, vol. 8, Issue 6, June 2018. https://doi.org/10.3390/geosciences8060205

[6] Villafuerte I., Barrazueta A. y Corral C., Desarrollo Turístico de la Ruta del Oro y su área de influencia en los cantones Zaruma y Portovelo, Tesis, 2005. (In Spanish)

[7] Oviedo B., El Patrimonio Industrial Minero del Archivo Histórico y Museo de Minería, Asociación Civil y la Carta de Nixhny Tagil. ICOMOS- México AC, 2015. (In Spanish)

[8] Paz B., Calderón B. y Ruiz H., La Gestión Territorial del Patrimonio Industrial en Castilla y León (España): fábricas y paisajes. *Investigaciones Geográficas, Boletín del Instituto de Geografía, UNAM*, para ser publicado. (In Spanish)

[9] Nuñez P., Vejsbjerg L., El Turismo entre la Actividad Económica y el Derecho Social El Parque Nacional Nahuel Huapi, Argentina, 1934–1955. *Estudios y Perspectivas en Turismo*, Universidad Nacional de Rio Negro- Argentina, vol. 19, pp. 930–945, 2010. (In Spanish)

[10] Castillo A., López T. y Vázquez G., El turismo industrial minero como motor de desarrollo en áreas geográficas en declive. Un estudio de caso. Estudios y Perspectivas en Turismo. Universidad de Córdoba-España, vol. 19, pp. 382–393, 2010. (In Spanish)

[11] Sandoval F., Albán J., Carvajal M., Chamorro C., y Pazmiño D., Minería, Minerales y Desarrollo Sustentable en Ecuador, *Ambiente y Sociedad*, Informe Nacional MMSD-Ecuador, cap. 7. (In Spanish)

[12] Domínguez I., Costa V. y Guardado R., La comunicación en el patrimonio geológico minero: un enfoque desde la minería del cromo en Moa. *Minería y Geología*, vol. 31 n.3, pp. 128–139, ISSN 19938012, julio-septiembre, 2015. (In Spanish)

[13] Montero J. & Salazar Y., La reinserción laboral tras el cierre de minas: una vía para lograr el desarrollo sustentable en la minería. *Minería y Geología*, vol. 27 n.4, pp. 64–87, ISSN 19938012, octubre-diciembre, 2011. (In Spanish)

[14] Fernández G. & Guzmán A., El patrimonio industrial-minero como recurso turístico cultural: El caso de un pueblo-fábrica en Argentina, *PASOS Revista de Turismo y Patrimonio Cultural*, vol. 2 N°. 1, pp. 101–109, 2004. https://doi.org/10.25145/j.pasos.2004.02.008. (In Spanish)

[15] Ministerio de Industrias y Productividad, Coordinación Zonal 7, Plan estratégico emergente de recuperación y fomento productivo de los cantones, diciembre, 2017. (In Spanish)

[16] Moscoso J., Política Pública de Reparación Integral. Estudio de caso en el Distrito Minero Zaruma-Portovelo, Provincia de El Oro, *Disertación previa a la obtención del título de sociólogo, con mención en desarrollo*, Pontificia Universidad Católica del Ecuador, 2015. (In Spanish)

[17] Valladares R. Y., Dal Pozzo F. y Castillo A. J., Propuesta metodológica para definir la vocación minera en el contexto del ordenamiento territorial venezolano. *Boletín Geológico y Minero*, 126 (4), pp. 663–676, ISSN 0366-0176, 2015. (In Spanish)

[18] Orche E., Parques mineros, desarrollo sostenible y ordenación del territorio. *Integración de la minería en la ordenación del territorio. ESPOL-CYTED-CICYT. Guayaquil*, pp. 112–124, 2003. (In Spanish)

[19] Orche E., Puesta en valor del patrimonio geológico-minero: el proceso de adaptación de explotaciones mineras a parques temáticos. *Patrimonio geológico y minero en el contexto del cierre de minas. CETEM-CYTED-IMAAC-CNPq. Río de Janeiro*, pp. 51–65, 2003. (In Spanish)

[20] Guerrero-Almeida, D., Chacón Pérez, Y., Fonseca Hernández, D. y Court-Potrillé, M. Metodología para la ejecución de un cierre de minas sustentable. Minería y Geología. Online https://www.redalyc.org/articulo.oa?id=223532481006, **30(3)**, pp. 85–103, 2014. Accessed on: 16 Aug. 2019. (In Spanish)

[21] Carrión P., Amos V., Ladines L., Loayza G., Domínguez M. J. y Berrezueta E. La ruta del Oro y el Patrimonio geológico-minero en Zaruma-Portovelo (Ecuador). *IV Congreso Internacional sobre Patrimonio Geológico y Minero*, S-2, C-09, pp. 333–346, Ultrillas- 2003. (In Spanish)

[22] Alberruche del Campo E., Marchán C., Sánchez A., Ponce D. y García A. Guía metodológica para la integración del Patrimonio Minero en la Evaluación de Impacto Ambiental. *Encomienda de gestión de trabajos en materia de impacto ambiental y de producción y consumo sostenible*, Instituto Geológico y Minero de España; Ministerio de Agricultura, Alimentación y Medio Ambiente, 40 págs., 2012. (In Spanish)

[23] National Secretary of Planning and Development and National Information System, Information for planning and territorial planning, National Geographic Atlas, 2013. http://sni.gob.ec/atlas-geografico-nacional-2013

[24] Murillo Carrión R. La Colonia: una época de transición. *Zaruma, historia minera Identidad en Portovelo*, cap. I, ed. BRYA-YALA. Docutech Quito-Ecuador, 2000. (In Spanish)

PLANNING LONG-TERM MANAGEMENT FOR HISTORIC CITIES: THE ROCK INTEGRATED AND SUSTAINABLE MANAGEMENT PLAN

ANDREA BOERI, DANILA LONGO, CHIARA MARIOTTI & ROSSELLA ROVERSI
Alma Mater Studiorum, University of Bologna, Department of Architecture.

ABSTRACT

The practice of management, although quite recent, embodies one of the major challenges for planning sustainable historic cities. Environmental, climatic, social and cultural issues affecting historic contexts, make management a fundamental process in balancing sustainable development and heritage conservation over time. In this framework, the ROCK – *Regeneration and Optimisation of Cultural heritage in creative and Knowledge cities* – project (H2020, n. 730280) aims at developing Cultural Heritage-led regeneration strategies capable of ensuring streams of long-lasting structured actions after the end of the project. Involving 10 European cities under the coordination of the Municipality of Bologna with the core scientific support of the University of Bologna, this project is designing a long-term management tool and testing its innovative pathway across the EU. Thus, this paper will focus on the genesis, construction and implementation of the ROCK Integrated Management Plan (IMP) as vital toolkit for developing, over time, sustainable urban procedures in the historic cities focused on the essential role of tangible and intangible heritage. Based on the UNESCO Management Plans, the existing literature and on several international reference experiences, ROCK IMP embeds the research-action-research methodology that underlies the whole project. Following this innovative circular approach, a preliminary set of goals and actions were selected (research), then implemented trough pilot interventions, co-designed and co-created with local administrations, stakeholders and citizens (action), and finally updated, considering the impacts of the performed actions to recalibrate the process and build scenarios for the IMP (research). As it is an ongoing process, this paper will appraise its state of the art, highlighting the potential and criticalities and describing what has been done in the city of Bologna so far.
Keywords: Cultural Heritage, governance, historic cities, integrated management, long-term planning, urban sustainable regeneration.

1 INTRODUCTION

'Urban decay, social conflict and low living standards are not uncommon in many of European historic city centres. Is it possible to breathe new life into these areas while doing it in a sustainable way?' [1].

This is the starting point of a very recent report published by the European Commission and focused on scientific research advances concerning the future of the historic built environment. A very relevant aspect for the present study is not only the affirmative statement of the European Commission on the matter, but the exemplary project presented for facing this challenge. This is the European project 'ROCK – *Regeneration and Optimisation of Cultural heritage in creative and Knowledge cities*' – (grant agreement n° 730280), funded under the Horizon 2020 Work Programme 2016–2017 *Climate action, environments, resource efficiency and raw materials*, topic *Cultural Heritage as a driver for sustainable growth* (SC5-21) as part of the call *Greening the Economy*, third Horizon thematic pillar on *Societal Challenge* [2].

ROCK develops innovation in Cultural Heritage (CH) through a collaborative and systemic approach that promotes CH-led regeneration strategies for historic city centres and

DOI: 10.2495/DNE-V14-N4-311–322

supports the demo areas' transformation into Creative, Cultural and Sustainable Districts. In ROCK vision, cultural heritage is not an inactive and burdensome factor but 'a unique and powerful engine of regeneration, sustainable development and economic growth for the whole city' [3]. Starting from this assumption, ROCK expressed the ambition to go over the three years of the project – whose end is scheduled in May 2020 – in order to ensure future streams of long-lasting structured actions based on its pilot experience. To meet such a goal, a specific tool for achieving outputs and outcomes of successful management is been tested and developed: the ROCK Integrated Management Plan (IMP).

As development without conservation cannot be sustainable, while conservation cannot succeed without development to sustain its efforts [4], an integrated and long-term system of urban governance is today increasingly urgent. From this point of view, ROCK IMP represents one of the main output and, at the same time, the most important legacy of this research. The relevance of the *research-action* implemented by ROCK has been confirmed during the high-level H2020 Conference *Innovation & Cultural Heritage* held in Brussels on 20th March 2018: the project has been invited to participate to the conference organised by the European Commission to celebrate the European Year of Cultural Heritage and has been presented as a significant model of innovation for sustainable urban development and management [5].

2 OBJECTIVE AND METHODOLOGY

The objective of this paper is to describe the process of genesis, construction and implementation of ROCK Integrated Management Plan. As ROCK is an ongoing project, potential, criticalities and still unresolved issues regarding the current state of the art of this work are outlined.

The aim is to identify some of the most promising orientations in the field of urban regeneration, planning long-term management for the 'Historic Urban Landscape' [6], understanding the evolving nature of today's cities and addressing it in a sustainable way. Working on the stratified context of the European cities means facing a whole range of social, economic, environmental and conservation criticalities (for instance physical decay, social conflicts, environmental pressures, economic crisis, lack of security, ineffective management of spaces and underuse of existing buildings), but also benefitting from the presence of tangible and intangible cultural heritage as a fundamental driver of continuity and progress. For this reason, ROCK can be consider part of an experimental sector of cultural heritage research that bets on the possibility to balance into a unique project the major challenge for the future of European historic cities: managing the systemic urban transformation between conservation and innovation.

This paper is structured in two core sections. The first one offers a general overview on the concept of 'management' with particular reference to recent guiding principles for managing cultural heritage developed by international institutions such as UNESCO, ICCROM, ICOMOS, IUNC to name but a few. The second section deepens the ROCK experience of management through the description of the Integrated Management Plan, especially the design process for its construction. This section also illustrates similarities and differences between ROCK IMPs and other Management Plans already codified by the existing literature – in particular UNESCO Management Plans. Last but not least, first experimentations in Bologna ROCK city are presented in order to illustrate a concrete example of implemented actions.

The study is based both on a desk research, especially for the first section, and on a direct research experience on the ROCK project, mainly with regard to the second one.

3 MANAGING CULTURAL HERITAGE COMPLEXITY

3.1 The concept of 'management'

The concept of 'management' is comparatively recent having its root in the *Convention concerning the Protection of the World Cultural and Natural Heritage* as 'the duty of ensuring the identification, protection, conservation, presentation and transmission to future generations of the cultural and natural heritage' [7]. Since then, the need to achieve these purposes has become more and more complex due to the increasing pressure on the contemporary world and to the widening definition of Cultural Heritage [8].

As far as the European polices for CH are concerned, the trend is to recommend 'a holistic research agenda and an inclusive interdisciplinary approach' [9], by means of a new vision and mission in heritage management. In this context, the historic built environment is defined as an inclusive and comprehensive platform that cannot be understood or managed except through an approach that embraces all its complexity.

The main aim is to raise understanding of the integration of urban and heritage planning in multilevel governance, exploring ways to best reveal the relations between supranational and subnational policy [4]. In this perspective, new targets have been established such as the importance of a common ground to define, assess and improve management systems, the mutual exchange of good practices and the evolution of improved management approaches as well as the provision of practical guidance and tools for day-to-day management practice recognizing the increased number of stakeholders involved and the awareness of the diversity of management problems linked to each specific country [10].

3.2 The UNESCO Management Plan

According to these key-assumptions, UNESCO, with the support of other international institutions, published *The World Heritage Resource Manual: Managing Cultural World Heritage*, whose aims is to understand the management systems and the ways to improve them for managing cultural heritage effectively. The *Manual* identifies nine basic components and clusters them in three interlinked pillars [10] (Fig. 1):

- 3 elements: legal framework, institutional framework and resources;
- 3 processes: planning, implementation and monitoring;
- 3 results: outcomes, outputs and improvements.

The Management System is the result of the combined effect of all these components. Its conceptual structure provides a common framework and helps heritage practitioners, policy-makers, communities, local stakeholders and citizens to control the systemic evolution of heritage processes by ensuring conservation and enhancement of its Outstanding Universal Value. As Outstanding Universal Value is meant 'cultural and/or natural significance which is so exceptional as to transcend national boundaries and to be of common importance for present and future generations of all humanity' [11].

The Management System is fully reflected in the Management Plan, intended as 'the guidance document developed within, and describing, a particular management system' [10]. As defined in Appendix A of the above-mentioned *Manual*, it is 'a relatively new tool which determines and establishes the appropriate strategy, objectives, actions and implementation

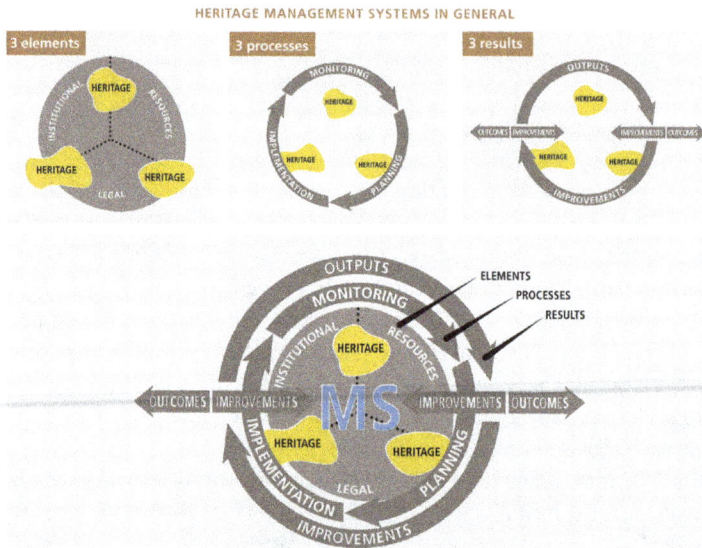

Figure 1: UNESCO Heritage Management System.
Source: UNESCO et al. 2013 [10].

structures to manage and, where appropriate, develop cultural heritage in an effective and sustainable way so that its values are retained for present and future use and appreciation. It balances and coordinates the cultural heritage needs with the needs of the 'users' of the heritage and the responsible governmental and/or private/community bodies' [10].

Therefore, a responsive Management Plan should be *unique* and *site-specific, cyclical* to evaluate its process so as to adjust its ongoing activities and to inform the next cycle, *interacting* with other Management Systems or with their components, *sufficiently flexible* to deal with unforeseeable events (e.g. natural disaster and financial changes), *common* and *shared* as it is shaped on inputs of the different actors involved, *dynamic* and *built through a participatory approach, drafted for the long-term future* of the property, *realistic* and *sustainable* to assure multi-level benefits respecting the past, taking advantage of the present and not damaging the future, and last but not least, *regularly reviewed* and *updated* in order to better respond to evolving challenges [10, 12].

Such principles have been translated into several Management Plans aiming at balancing conservation and sustainable development as well as at safeguarding the Outstanding Universal Value, the authenticity and integrity of all sites included in the *UNESCO World Heritage Lists*. A still open challenge is the opportunity to transfer and extend this model to wider contexts, for instance to historic cities across Europe. The ROCK project is founded on this challenge.

4 PLANNING INTEGRATED MANAGEMENT FOR HISTORIC CITIES

4.1 The H2020-ROCK project

ROCK is the European project funded under the Horizon 2020 Programme in 2017. ROCK stands for *Regeneration and Optimisation of Cultural heritage in creative and Knowledge cities*. Its mission supports one of the four core actions recommended by the European Union (EU) to 'reinforce the role of heritage as part of Europe's underlying socio-economic, cultural and natural capital', that is, *Heritage Led Urban Regeneration* [13]. This project has received funding of about 10 million euros and currently involves 32 partners from 13 European countries and 10 historic cities of the Union, coordinated by the Municipality of Bologna in close scientific collaboration with the Alma Mater Studiorum University.

ROCK focuses on historic city centres as 'extraordinary laboratories' where testing systemic transformations into Creative, Cultural and Sustainable Districts through shared generation of new sustainable environmental, social, economic processes [3]. Starting from a selection of three demonstrative areas in Bologna, Lisbon and Skopje, the project aims at overcoming the physical decay, social conflicts and poor life quality affecting these areas, and at promoting strategies to spread knowledge, experience and good practices in order to help municipality leaders and the local ecosystem of stakeholders in developing the vision, and gaining the skills, to be successful at using heritage to regenerate their cities [3, 13].

This project is based on the mutual exchange of meaningful experiences of socio-cultural transformations between the 10 cities involved: 7 *Role Model Cities* (Athens, Cluj-Napoca, Eindhoven, Liverpool, Lyon, Turin and Vilnius), which have already implemented vital processes of regeneration, and 3 *Replicator Cities* (Bologna, Lisbon and Skopje) where similar successful models can be transferred in relation to their local contexts.

ROCK pilot process is driven by a *research-action-research* methodology that marks the essential role of the research in heritage development procedures: the idea is to implement site-specific actions, record their feedbacks, verify their impacts, and recalibrate them for the future, again with the support of the research.

The pilot process implementation is structured around three main pillars: *accessibility, sustainability* and *new collaborations for productions*; within each of them, detailed thematic actions have been clustered (e.g. lightings, wayfinding, greening, business matching, adaptive reuse, knowledge building and so on). In addition, key components of the project are the enabling technologies – especially Information and Communication Technologies for cultural heritage promotion and dissemination such as Virtual and Augmented Reality – and the attention to the environment, complementary to both the social and the conservation ones.

4.2 Towards ROCK Integrated Management Plan

On 20th May 2014 the Council of the European Union adopted a fundamental document, which has recognized Cultural Heritage as 'a strategic resource for a sustainable Europe' [14]. In line with the EU *Conclusions* and as already stated, ROCK develops innovation in the field of cultural heritage management through the construction of a key tool for planning long-term Cultural Heritage-led regeneration strategies in historic city centres.

The ROCK Integrated Management Plan is designed to ensure the right balance between cultural heritage conservation and urban sustainable growth in a long-term perspective. For this reason, it is intended as a very practical, easy-to-understand and easy-to-use toolkit, and it is based on the existing literature as well as on virtuous management experiences already implemented. The UNESCO Management Plan represents its main reference point, especially those drafted by the Role Model Cities: Liverpool, Lyon and Vilnius [15–17]. After a preliminary phase of knowledge building, ROCK adopts principles and structure of UNESCO Plans while introducing two main innovative aspects.

The first innovation regards the use of a new attribute. Compared to the UNESCO Management Plan, the ROCK one is called *Integrated* because it supports a multi-level integration process:

- it integrates the existing city plans and procedures without being an additional governance tool;
- it integrates all relevant categories of users interested in urban policies and CH research, thus facilitating trans-disciplinary and trans-sectorial collaborations;
- and it continuously integrates research and action phases.

The second innovation coming from ROCK is directly linked to this latter aspect as the result of the *research-action-research* methodology. In fact, the design process of ROCK Integrated Management Plan is based on this circular approach including:

- a first *research phase* aimed at identifying needs, key stakeholders, key areas and key actions and enablers, in order to properly prepare the concrete action planning and implementation. This preliminary set of actions helps cities to build scenarios to predict and simulate different possibilities of CH-led sustainable growth with regard to the three main pillars of the project (accessibility, sustainability and new collaboration). A participatory approach chains the co-designed activities, deriving from the mutual exchange between Role Model and Replicator cities, as well as the Living Lab meetings;
- an *action phase,* focused on the implementation of pilot actions according to the first draft of scenarios assumed by Replicators and supported by a mixed bottom-up and top-down initiatives. The pilot implementation can meet constraints and barriers, make arise more needs, problems and manifest the necessity of adjustments. Therefore, the process foresees additional research activities;
- a second *research phase,* finalized to the definition of more detailed scenarios capable of taking into account new assumptions and needs deriving from the implemented pilot actions.

ROCK Integrated Management Plan grafts itself on this second research phase: hence, the scenario construction is the preliminary and crucial work that supports cities in defining their IMP. The major aim of ROCK IMPs is to facilitate the transition from the experimental activities carried out during the project to a long-term structured action programme for the future. In order to achieve this objective, it is composed of two complementary parts:

- an *Operating Manual* (one for all Replicators) including the methodology and the common structure underlining ROCK Integrated Management Plan – knowledge gain, vision definition, action plan, future perspectives –;
- and three *site-specific Management Plans* (one for each Replicator) to be compiled during the life time of the H2020 project and delivered at the end of the entire research experience.

According to these principles, ROCK Integrated Management Plan launches a process founded on a horizontal integration of actors – a mix of planned and emergent elements, self-organized activities coalescing into a shared model of local development, and a cross-sectoral coordination of polices, plans and procedures for the historic site implementation. Such a methodology is meant to be applied at the urban scale, nevertheless ROCK proposes a first application in smaller and more specific contexts: pilot actions are carried out at the district scale that offers an intermediate dimension useful for supporting projects and monitoring results effectively [18].

At present, the three Replicator Cities are working synergistically to build their respective Management Plans from which to derive the contents of the Operating Manual. To facilitate the construction phase, which is still in progress, some practical factsheets and several collective meetings have been developed for guiding, step by step, the actors involved in the implementation and integration process of this new model of urban governance.

As a matter of fact, the Integrated Management Plan mainly supports the ROCK regeneration whose aim is to activate a new, systemic and circular flow of urban transformations. As shown in the graphic concept (Fig. 2), it turns around the predictive macro-scenarios on accessibility, sustainability and new collaborations for production that really make real the revitalization of spaces and relationships into historic cities.

Conceptually, each scenario can be traced back to a series of universal guiding principles that have a different degree of relevance in different contexts and that must be declined according to each local site.

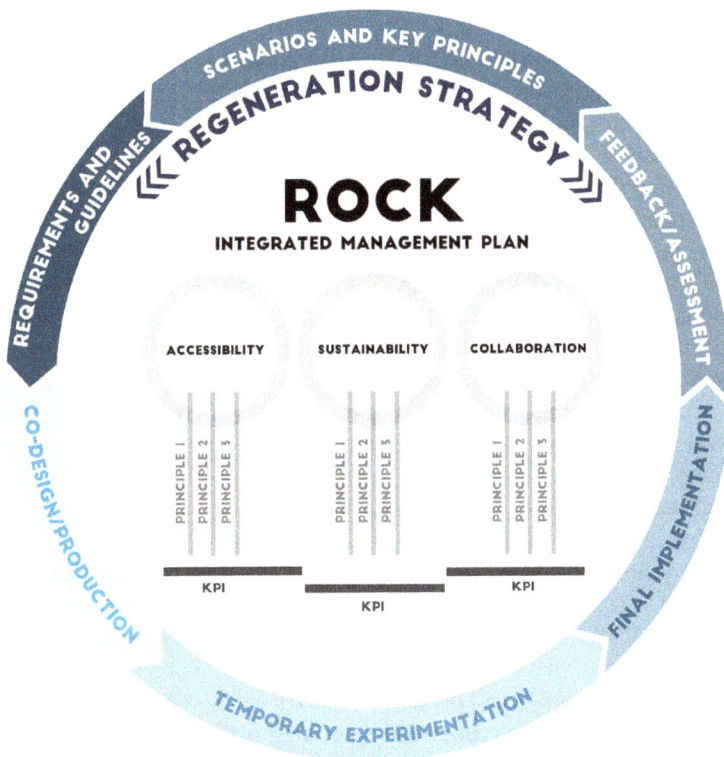

Figure 2: The ROCK Integrated Management Plan: design and implementation process.

Source: Graphic design by Chiara Mariotti.

The evaluation process regarding the significance of scenarios and their principles, as well as the selection and implementation of regeneration pilot actions, is carried out with a 'living lab' approach and will be checked, monitored and updated thanks to the support of suitable Key Performance Indicators.

Therefore, the ROCK strategic regeneration process starts after the scenario construction and the guiding principles identification, and consists of the following circular steps:

- requirements definition and guidelines assumption to orient activities with clear objectives and rules;
- co-design and co-production activities for supporting a multi-actors process;
- temporary and final experimentation to give effect to urban transformations;
- feedback system so as to assess in a quantitative and qualitative way the response to implemented actions.

Through the description of the ROCK IMP design process, this paper wants to transmit the complexity of long-lasting social, economic, environmental and cultural dynamics activated by the project. The following section gives a more concrete perspective describing what is especially being done in Bologna.

4.3 First applications in Bologna ROCK city

The experience of Bologna is one of the most interesting because two Management Plans are currently under construction: the ROCK Integrated Management Plan, as the main output of H2020 research activities, and the UNESCO Management Plan, as the key document supporting city porticoes' candidacy into the *UNESCO World Heritage Lists*.

The Bologna demo site is the University area (called *U-Zone* by ROCK) located inside the historic city centre and developed over time along Via Zamboni, one of the major street characterized by the presence of typical Bolognese porticos and urban squares (Piazza Ardigò, Piazza Rossini, Piazza Verdi, Piazza Scaravilli e Piazza Puntoni), and rich in cultural and artistic institutions and museums. Nevertheless, this cultural heritage is not fully known and exploited by citizens and tourists and frequent social conflicts have arisen from the forced coexistence between residents and student population.

The selected pilot actions are the result of a process of community involvement (named *U-Lab*) to create a local Ecosystem of Stakeholders and to enable the co-designed and co-construction process (*U-Atelier*). According to the IMP structure, a first diagnostic phase aimed to underline constraints at different levels has been addressed, followed by the identification of needs and priorities that can boost actions implementation.

During the two years of ROCK project, different types of actions have been carried out in the U-Zone, each one inserted in a global vision, concurring to achieve integrated targets and objectives and according with the vocation of single spaces (indoor and outdoor, public and private).

New lightscape is one of the already implemented actions involving Via Zamboni and its surrounding areas. For several years, phenomena of social degradation and physical decay have been reducing the sense of security and restricting use and fruition of the area, especially in the night. During thematic participative tables and on-site meetings, involving also disability manager and associations of people with disabilities, the need to rethink lighting along the historical axis of Via Zamboni strongly emerged as possible strategy to improve accessibility (Fig. 3). Thus, *new lightscape* has been conceived by Viabizzuno – partner of the

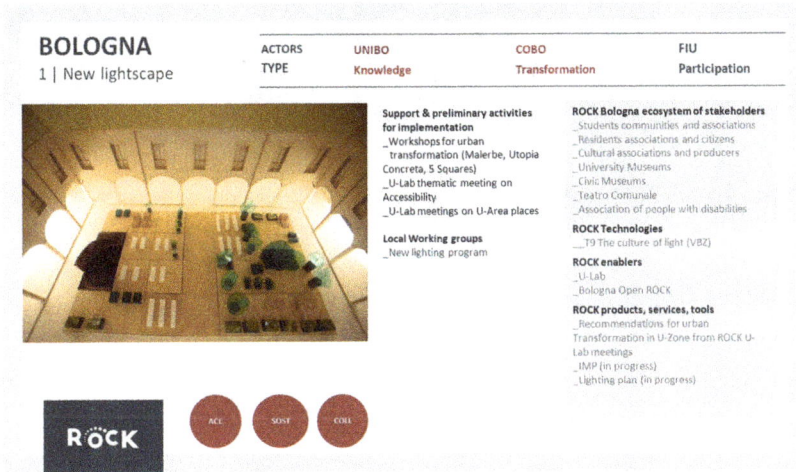

Figure 3: New lightscape overview table.
Source: U-lab and U-Atelier.

ROCK project – as an experimentation of innovative lighting concept that tests effective solutions to combine physical accessibility and security of public spaces with cultural heritage enhancement. The lighting project intends to valorise the historic setting through the adoption of specific solutions designed for the peculiar configuration of spaces, environmental conditions and features of buildings. Light contributes to a better perception of volumes, distinction between voids and solids, colours, movements. Besides, the light is conceived as a detector of hidden treasures and points out their presence, treating the city as an art gallery, for example showcasing significant architectural details (Fig. 4).

The project takes into account the variability of the use of spaces and kinds of users in relation with the different times of fruition, modulating the lighting system into: light to meet (from dusk until 19:30), light for break (from 19.30 to 20.30), light for architecture and art (from 20:30 to 23:30), and light for sleeping (from 23:30 until sunrise). Therefore, the developed guiding criteria closely link the use of light to the fruition of public space, intervening on the environment as a system of structural interrelationships between an individual and its relevant space.

Figure 4: Lighting project for Piazza Rossini and its simulation.
Source: Viabizzuno.

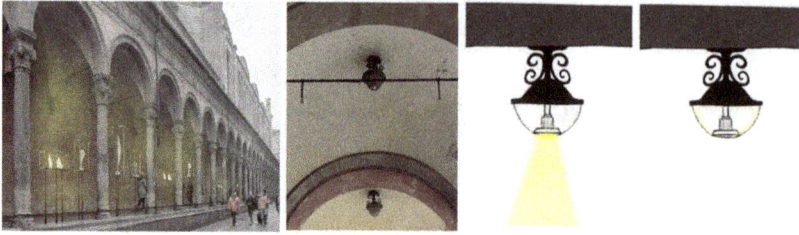

Figure 5: Via Zamboni porticos: display-cases for temporary exhibitions and historical lanterns subjected to conservative restoration.

Source: Viabizzuno.

The project includes the restoration of the historical lanterns under the porticoes and the conservation of the old notice boards, still affixed to the walls, with the addition of integrated lighting to show their contents clearly. New display-cases for art works and temporary exhibitions have been also designed, resistant to vandalism thanks to a base fixed to the floor and characterized by a downward light tracing the path and lighting the surrounding area; bicycle carriers with integrated light improve safety as well as information totems helps people get historical information and orient themselves (Fig. 5). The project integrates technical competences based on the research of Viabizzuno. All the designed solutions concur to literally put the area under a new light: improve physical accessibility, increase safety, enhance cultural heritage, encourage public events, support new forms of sociality and inclusion, and promote a new urban storytelling.

The positive outcome of these urban transformations confirms the role of lighting as key action to focus on in order to cluster a set of vital experimentations, which can guide the action planning to be included in Bologna ROCK Integrated Management Plan.

5 CONCLUSION AND OPEN ISSUES

This paper has described the ROCK experimentation on cultural heritage-led regeneration strategies for historic sustainable cities. The heart of the investigation has focused on the design and implementation phases of a new governance tool, the Integrated Management Plan, capable of supporting a multi-level integration – plans, procedures, policies, actors and processes – and harmonizing cultural heritage conservation and enhancement, and sustainable development in a long-term perspective.

As ROCK is an experimental research and the construction of the Integrated Management Plan is still ongoing, it is necessary to point out that, on one hand, many of the addressed questions are still evolving and that, on the other hand, there is no lack of open issues. Nevertheless, the ROCK project has several strong points, such as the ability to broaden participation in planning long-term management for historic cities. From this point of view, the importance of the 'democratic participation' and the creation of 'heritage communities', both recommended by the *Framework Convention on the Value of Cultural Heritage for Society* (Faro Convention), is strongly confirmed [19]. Another crucial factor is the concept of 'integration' – expressed also in the name of this tool – intended as the need to promote a governance instrument capable of activating new social, economic and cultural dynamics, strongly rooted in each urban local context. In this regard, Bologna is a far-sighted example: the city candidacy for porticoes into the *UNESCO World Heritage Lists* is in fact benefitting from the Integrated Management Plan put in place by ROCK and vice-versa, in an innovative perspective that recognizes the transversal value of cultural heritage management for future generations. The Municipality of Bologna is working together with the Research Group of the University of Bologna to draft this plan, whose guidelines will be disseminated at the end of the design process. The on-site actions allow to test strategies and to point out constraints, barriers and unexpected positive implications, as happened thanks to the described *New lightscape*, which highlighted the opportunities to preserve cultural heritage and enhance its accessibility by working on immaterial aspects.

In conclusion, the main innovative contribution coming from ROCK project is the *Roadmap for sustainable Cultural Heritage Management in historic cities*. To achieve this goal, some open issues, specifically related to ROCK Integrated Management Plan, have to be deepened and solved, such as the authorship, duration and territorial area of the plan as well as the strategy to keep it alive after the end of the project; these are all essential queries that will be addressed in the ongoing phases of the project.

ACKNOWLEDGEMENTS

ROCK project is funded by European Union's Horizon 2020 Research and Innovation Programme (call H2020-SC5-2016-2017, Grant Agreement n. 730280).

REFERENCES

[1] European Commission, Using culture to breathe new life into historic city centres, 18 June 2019. Online, https://cordis.europa.eu/news/rcn/131357/en (accessed 01 July 2019.)

[2] European Commission, Horizon 2020, Work Programme 2016–2017 – 12. Climate action, environment, resource efficiency and raw materials, 24 April 2017. Online, https://ec.europa.eu/research/participants/data/ref/h2020/wp/2016_2017/main/h2020-wp1617-climate_en.pdf (accessed 01 July 2019).

[3] ROCK Official Website. Online, https://rockproject.eu (accessed 01 July 2019).

[4] Veldpaus, L., *Historic Urban Landscapes: Framing the Integration of Urban and Heritage Planning in Multilevel Governance*, Technische Universiteit Eindhoven: Eindhoven, 2015.

[5] European Commission, Horizon 2020 Cultural Heritage and European Identities. List of projects 2014–2017. Directorate-General for Research and Innovation Open and inclusive Societies, 2018. Online, https://ec.europa.eu/research/social-sciences/pdf/project_synopses/cultural_heritage_projects.pdf (accessed 01 July 2019).

[6] UNESCO, Recommendation on the Historic Urban Landscape, Paris, 10 November 2011. Online, https://whc.unesco.org/uploads/activities/documents/activity-638-98.pdf (accessed 01 July 2019).

[7] UNESCO, Convention concerning the Protection of the World Cultural and Natural Heritage, 1972. Online, https://whc.unesco.org/en/conventiontext/ (accessed 04 July 2019).

[8] Jokilehto, J., Definition of Cultural Heritage. References to Documents in History. ICCROM Working Group "Heritage and Society", 2005. Online, http://cif.icomos.org/pdf_docs/Documents%20on%20line/Heritage%20definitions.pdf (accessed 04 July 2019).

[9] Sonkoly, G. & Vahtikari, T., Innovation in Cultural Heritage Research. For an integrated European Research Policy, Directorate-General for Research and Innovation Europe in a changing world – Inclusive, innovative and reflective societies (Horizon 2020/SC6)

and Cooperation Work Programme: Socio-Economic Sciences and Humanities (FP7), Publications Office of the European Union in Luxembourg, 2018. doi: 10.2777/673069. Online, https://publications.europa.eu/en/publication-detail/-/publication/1dd62bd1-2216-11e8-ac73-01aa75ed71a1(accessed 04 July 2019).

[10] UNESCO, ICCROM, ICOMOS, IUNC, Managing Cultural World Heritage, UNESCO World Heritage Centre – World Heritage Resource Manual: Paris, 2013. Online, https://whc.unesco.org/document/125839 (accessed 04 July 2019).

[11] UNESCO, Operational Guidelines for the Implementation of the World Heritage Convention, UNESCO World heritage Centre: Paris, 2017. Online, https://whc.unesco.org/en/guidelines/ (accessed 04 July 2019).

[12] UNESCO, Management planning of the UNESCO World Heritage Sites. Guidelines for the development, implementation and monitoring of management plans. With the examples of Adriatic WHS, Expeditio: Kotor, 2016. Online, https://www.expoaus.org/upload/novosti/publication_expoaus_eng_web_105355.pdf (accessed 04 July 2019).

[13] European Commission, Getting cultural heritage to work for Europe. Report on the Horizon 2020 Expert Group on Cultural Heritage, Directorate-General fro Research and Innovation, Publications Office of the European Union: Luxembourg, 2015. Online, https://www.kowi.de/Portaldata/2/Resources/horizon2020/coop/H2020-Report-Expert-Group-Cultural-Heritage.pdf (accessed 05 July 2019).

[14] Council of the European Union, Conclusions on cultural heritage as a strategic resource for a sustainable Europe, Brussels, 20 May 2014. Online, https://www.consilium.europa.eu/uedocs/cms_data/docs/pressdata/en/educ/142705.pdf (accessed 05 July 2019).

[15] LOCUS Consulting Ltd, Liverpool Maritime Mercantile City World Heritage Site Management Plan 2017–2024, 2017. Online, http://regeneratingliverpool.com/wp-content/uploads/2017/07/PMD-486-Liverpool-WHS-Management-Plan-FINAL-VER-SION-as-at-12-May-2017.pdf (accessed 06 July 2019).

[16] Mission site historique de Lyon - Direction des affaires culturelles, Plan de Gestion du site historique de Lyon. Inscrit sur la liste du patrimoine mondial de l'UNESCO, 2013. Online, http://whc.unesco.org/fr/documents/138552/ (accessed 06 July 2019).

[17] HERO – HERitage as Opportunity, Local Action Plan. Vilnius Old Town, 2011. Online: https://urbact.eu/sites/default/files/vilnius_action_plan.pdf (accessed 06 July 2019).

[18] Gaspari, J., Boulanger, S.O.M. & Antonini, E., Multi-Layered design strategies to adopt smart districts as urban regeneration enablers. *In International Journal of Sustainable Development and Planning*, **12(8),** pp. 1247–1259, 2017. Online, https://www.witpress.com/Secure/ejournals/papers/SDP120801f.pdf (accessed 06 July 2019).

[19] Council of Europe, Framework Convention on the Value of Cultural Heritage for Society, Faro, 27 October 2005. Online, https://rm.coe.int/1680083746 (accessed 07 July 2019).

ENERGY REHABILITATION OF BUILDINGS THROUGH 20 PHASE CHANGE MATERIALS AND CERAMIC VENTILATED FAÇADES

VÍCTOR ECHARRI IRIBARREN[1], JOSÉ L. SANJUAN PALERMO[2], FRANCISCO J. ALDEA CASTELLÓ[2] & CARLOS RIZO MAESTRE[1]
[1] Department of Building Construction, University of Alicante, Spain.
[2] University of Alicante, Spain.

ABSTRACT
In recent years, phase change materials (PCMs) have gained major relevance for their ability to take advantage of indoor/outdoor air temperature differences to store energy. This characteristic of PCMs allows to transfer stored energy to periods of energy demand, thus achieving optimum conditions of comfort and notable energy savings. The present study compared the energy consumption of a traditional façade and a ventilated façade to which large format ceramic tiles covered with PCMs were applied. For this purpose, an office building in the city of Alicante was used as a case study. Salt hydrate PCMs were attached to the slabs, and air was allowed to circulate or not circulate through night and day dampers as passive conditioning, accumulating energy. The energy performance of the building was simulated using the Lider-Calener (HULC) energy certification tool in both scenarios. The building's energy demand was calculated in its current state and with the ventilated façade with ceramic tiles and PCMs. An energy saving of 5% was obtained.
Keywords: ceramic, energy rehabilitation, phase change materials (PCM), ventilated façades.

1 INTRODUCTION
The housing sector is one of the European Union's major energy consumers: 40% of the energy produced in Europe is consumed throughout buildings' life cycle phases. For this reason, in 2010 the European Parliament approved Directive 2010/31/EU to reduce energy consumption and emissions in buildings [1]. Conforming with these European guidelines, Spain's Technical Building Code (or CTE by its Spanish acronym) was approved in 2006 to improve buildings' energy efficiency and consumption [2]–[3]. The building energy certification protocol based on the use of HULC software was approved at the same time [4].

Under these circumstances, it is necessary to consider the impact that the economic crisis and the bursting of the housing bubble have had on the Spain's construction sector. These factors have led to a drastic reduction in building construction activity, which is why many buildings are outdated. One way the governments stimulate the economy is by promoting building refurbishments, especially in the field of energy and consumption reduction. Thus, within this context of new demands for energy reduction, the present study proposes two lines of action. On the one hand, we suggest addressing energy losses and reducing consumption by intervening in façade enclosures to improve their performance by replacing joineries and by incorporating ventilated façade systems that prevent thermal bridges [5]. On the other hand, current needs in architecture to reduce energy consumption have generated new strategies based on passive energy conditioning systems in buildings. This reduces energy consumption and limits the use of traditional fossil fuels [6].

In recent years, thermal energy storage systems using phase change materials (PCM) have attracted considerable interest, since they allow adapting supply periods to energy demands. These systems present a great potential for improving energy efficiency [7]. The amount of publications on the subject of thermal storage using PCM has thus grown significantly [8].

© 2020 WIT Press, www.witpress.com
DOI: 10.2495/EQ-V4-N4-332-342

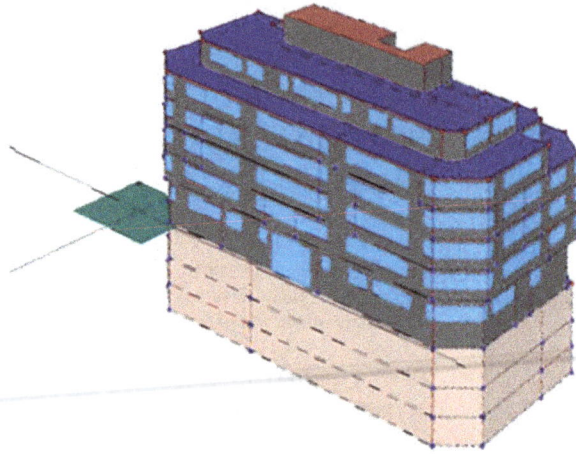

Figure 1: Model of the building drawn up with HULC software.

According to the statistics of the International Energy Agency (IEA), around 30% of energy supply is lost during its conversion, that is, due to dissipation or thermal costs [9].

This article focuses on the advantages and disadvantages of integrating and applying PCMs to buildings' energy refurbishment. A case study was conducted in Alicante city, in which annual energy demand was compared before and after incorporating new PCM construction systems. To quantify the improvements, both building models, the theoretical model and the real model, were simulated using the HULC software tool. The energy refurbishment proposal consisted in laying out the PCM on the lower face of the slabs, increasing their thermal inertia in the chamber formed by the false ceilings; controlled ventilation was added, allowing to regulate the flow of air entering from the outside. A new ventilated façade system with a porcelain stoneware ceramic finish was also incorporated.

2 VENTILATED FAÇADE

The ventilated façade with external thermal insulation is a construction system developed in northern European countries to solve certain construction problems [10]. The system creates a continuous barrier that protects the thermal envelope and prevents thermal bridges that cause energy losses in the building. This type of façade enables reducing energy demands and improving comfort levels [9]. Due to its characteristics, the system has also gained importance in parts of southern Europe such as Alicante.

A key component of these systems is the exterior finishing material, which needs to meet essential characteristics such as: strength, hardness, durability, impermeability and low weight, since it is difficult and expensive to maintain. Ceramic material is thus ideal for these types of façades due to its resistance, great durability and competitive cost [5].

3 PHASE CHANGE MATERIALS (PCM)

Passive energy storage systems such as those that capture solar radiation through greenhouse effects, Canadian wells that take advantage of land energy, or thermal gaps of outdoor air over night and day cycles, are some examples of how buildings' energy efficiency can be greatly improved. PCMs constitute an energy storage system, bringing thermal inertia to construction elements. These materials are capable of accumulating energy in the form of latent heat under given conditions of air temperature, solar radiation, etc. It is relevant to apply them to architectural solutions because of the energy savings that can be derived from them. When these materials are incorporated into building envelopes, they allow to better condition indoor spaces via radiant surfaces. Thus, they provide a high level of comfort and bring about substantial reductions in annual energy demands [7].

3.1 Introduction to PCM

PCMs are materials that undergo a change of state (liquid-solid-gas) at a given temperature. The amount of heat necessary to increase the temperature of one of these materials (sensible heat) by one degree is much lower than that required in the case of latent heat. The changes produced in the different materials due to latent heat occur at a given temperature that is proper to each material [6].

The interest in this type of material lies in the fact that during the phase change, the temperature remains constant while the material absorbs energy. This means that these materials have a higher energy density than other materials. Phase change materials that pass from solid to liquid require the least amount of energy [11]. In addition, theirs volumes undergo smaller variations than in other phase change processes. This allows using them in architecture in different applications and formats.

3.2 Types of phase change materials

The most common liquid-solid phase change materials used in the temperature range of 20°C to 80°C are: paraffin waxes, hydrated salts, eutectic mixtures and fatty acids.

- Paraffin waxes are available for commercial use. But their latent heat (approximately 200 kJ/kg) is only half that of hydrated salts, making them less interesting to use.

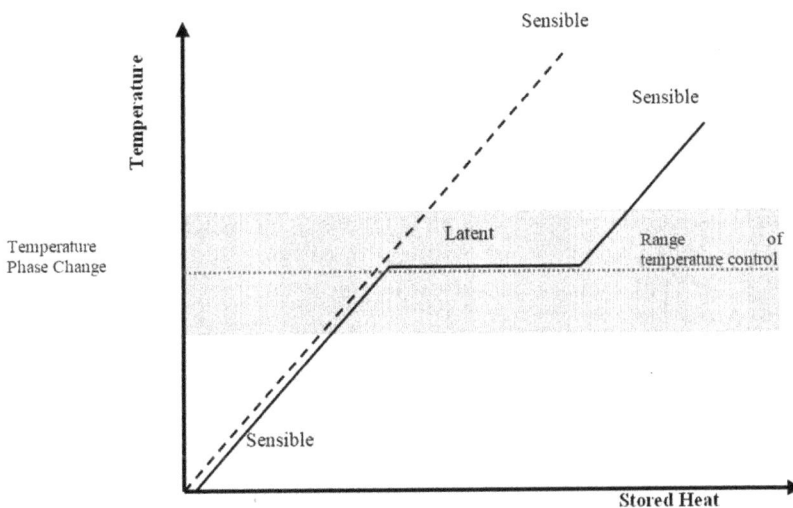

Figure 2: Operation process chart of phase change materials.

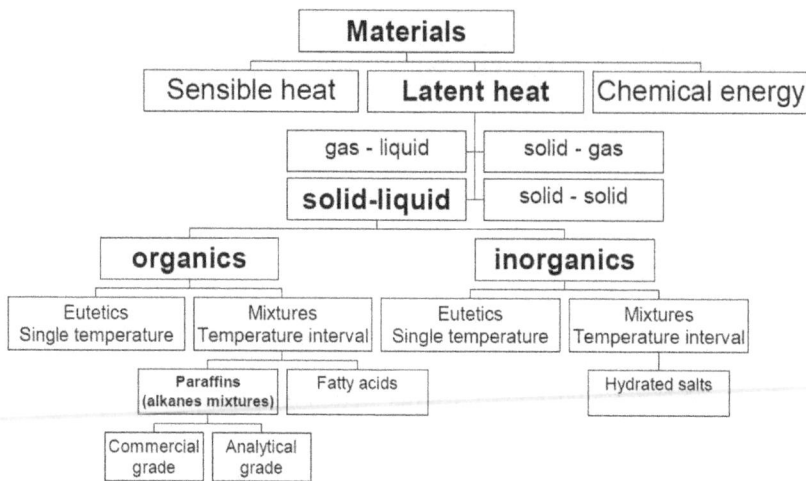

Figure 3: Classification of substances used for thermal storage.

- Hydrated salts are cheaper than paraffin waxes. They have the disadvantage of offering a low melting temperature and presenting higher corrosion when in contact with metals.
- Eutectic mixtures are formed by two components that, when joined, have a melting point (solidification) that is below that of the compounds individually, causing both elements to solidify at eutectic temperature. An example of these eutectic mixtures is salt bonding with ice.
- Fatty acids, like paraffin waxes, have a latent heat value of approximately 200 kJ/kg, but a higher commercial cost.
- Organic compounds do not present subcooling problems and are more stable than inorganic compounds [11]. Organic materials such as: waxes, fats and their esters have been recommended as heat accumulators, since their latent heat of fusion is 120 kJ/kg, their density 800 kg/m^3, their thermal conductivity 0.20 W/m·K and the specific heat is from 1500 J/kg·K.

3.3 Final considerations regarding PCMs

The use of PCMs for energy storage has increased rapidly in recent years. New products based on PCMs have appeared in the field of architecture. Applications of these products have been proven to be viable in other areas. New PCMs are being developed with determined characteristics and physical properties suitable for specific applications such as the temperature control of pharmaceutical products or of the human body [6].

Several studies have demonstrated how radiant floor heating and passive cooling by convection using energy storage systems based on PCMs function correctly in the construction field [9]. Results obtained by researchers on PCMs in a Mediterranean environment led us to opt for inorganic hydrated salts as they are inexpensive and offer a large heat storage capacity per volume unit [8]. Although organic substances are more stable and their temperature is closer to operating temperatures, they have the setback of presenting low thermal conductivity [11].

Figure 4: Paraffin wallboard [12].

Figure 5: PCM used in this work, hydrated salts SP 21EK by Rubitherm (Source: Rubitherm).

4 CASE STUDY

The headquarters of the National Organization for the Blind in Spain (ONCE by its Spanish acronym) in Alicante was chosen as a case study. The building uses a ventilated façade. The construction system is unique among the construction techniques prevailing in the area, which mostly use a traditional double-skin façade system with a chamber. Figure 7 illustrates

Figure 6: Image of the building.

the absence of thermal bridges in the façade thanks to the ventilated façade system, but also the presence of heat loss due to joineries with no thermal breaks.

The case study building is located in a widened area with street alignment, at 43, Avenida de Aguilera in Alicante city. The building was built in the year 2000 and consists in a seven-storey isolated block. It has four main façades with chamfered corners, resulting in eight façade planes, and is mainly used as an office building.

4.1 Composition of the existing enclosures

The exterior façade enclosure, from the exterior to the interior, is composed of: a finishing layer of natural granite stone plates of e = 2.5 cm, a ventilated air chamber of e = 6.0 cm, a galvanised steel metal substructure, an insulation thermal layer e = 4.0 cm, a plastering of cement mortar e = 1.5 cm, an interior sheet formed by hollow ceramic double brick e = 9.0 cm, a plastering and plaster finish of e = 1.5cm and an acrylic paint coating.

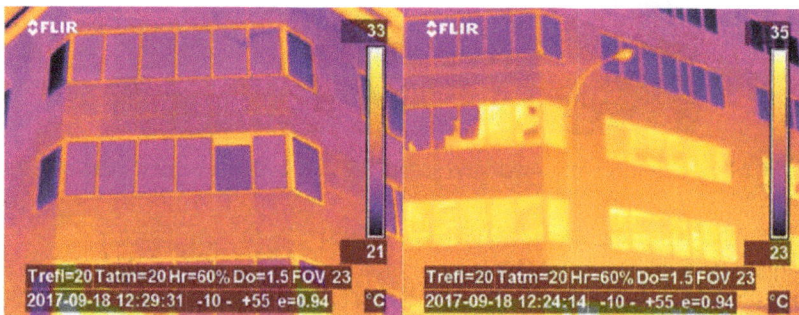

Figure 7: Thermal images of the ventilated façade. 18 September at 12:30 p.m.

5 ANALYSIS AND RESULTS OF THE PROPOSED SOLUTION

5.1 Description of the alternative solution

This work aims at innovating in the field of construction solutions to achieve the proposed objectives: passive conditioning, a ventilated façade using ceramic material and energy improvement. The general design was based on taking advantage of the PCMs' energy storage capacity and integrating the PCMs into the case study's joint refurbishment solution, arranging these materials in the false ceiling beneath each floor's slabs. In addition, new reflective thermal insulation was incorporated, improving the enclosure's performance, while decreasing thermal transmittance thanks to a thickness of only 3.0 cm (Fig. 8) [14]. We suggest replacing the existing aluminium window frames with high thermal transmittance by aluminium frames with a thermal bridge break based on a double-glazed sheet with an inner chamber since, as illustrated in the thermographs (Fig. 7), they are critical regarding energy loss.

5.2 Description of the passive conditioning system using PCM

A chamber communicating the building's opposite façades was proposed: it acts as an internal ventilated conduit in the false ceilings. Automated doors were placed on the façades to control the opening and closing of this conduit. The PCM sheets were placed in this chamber and acted as energy storage. The accumulated energy is gradually released to the upper slab in the form of heat both by conduction and by radiation [14]. The theoretical operation of the PCM conditioning system is detailed below based on four different scenarios.

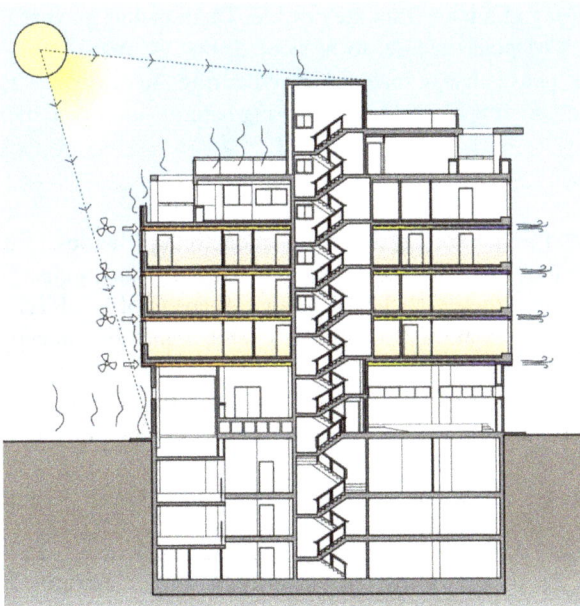

Figure 8: Comparative construction detail of the existing façade solution and the proposal using PCM.

5.3 System operation throughout the year

5.3.1 Winter

In the winter regime, an operating air temperature of 21°C [15], was taken into account inside the building, as Alicante city's maximum and minimum temperatures recorded in the winter period are 15.9°C and 6.0°C, respectively [16]. In addition, two cases were differentiated: diurnal and nocturnal cycles. During the day in winter, the system's PCM sheets capture outside air heat. This air accesses the chamber through the openings in the south façade. This façade is heated by solar radiation, raising the temperature of the air around it. In this case, both chamber doors are opened, those facing North and those facing South. The north façade openings may be closed to keep the hot air longer inside the chamber. The fans placed in the south façade can also be activated (if necessary) to improve the ventilation in the false ceilings. The PCM progressively increases in temperature, storing it in the form of energy [7]. The PCM plates heat the lower face of the floor by radiation and the air by convection when the system requires it. Under nocturnal winter conditions, the PCM plates gradually transfer the energy stored during the day in the form of heat to the lower face of the floor [17], the slabs acting as a radiant floor. This heat transfer takes place directly by conduction and indirectly by radiation [8]. The doors of the chamber must remain closed on both façades during the process.

5.3.2 Summer

In summer, an operating indoor air temperature of 23°C was taken into account as Alicante city's average maximum and minimum temperatures registered in summer are 31.6°C and 20.6°C, respectively. Two scenarios were considered: the day and night cycles [16]. During the day in summer, interior room temperatures are reduced, creating a flow of energy through the floor stored by the PCM via radiation (PCM-slab). For this, the doors are kept shut during the day, to avoid letting in hot air from the outside. There is thus no ventilation in the chamber. At night, the PCM plates are set to be cooled, that is, energy is to be transferred by convection from the phase change material stored during the day to the air [17]. To do this, the false ceiling chamber doors are opened creating natural ventilation during the night.

5.4 Façade proposal

The proposed solution (Fig. 8) brings an improvement to the joineries, as it integrates hinged joineries with thermal bridge breaks and double glazing with an air chamber. New reflective thermal insulation, 3 cm thick MULTITHERMIC 19 layers of WÜRTH® is proposed. The reduced insulation thickness allows for a bigger chamber and consequently, a larger air flow can circulate.

The major energy improvement measure to reduce annual energy demand consists in reducing U thermal transmittance of the enclosures making up the building's thermal

Table 1: Interior design conditions according to RITE (Spanish Regulation of Thermal Installations in buildings).

Season	Operating temperature °C	Relative humidity %
Summer	23–25	45–60
Winter	21–23	40–50

Table 2: Values of U thermal transmittance according to HULC (elaborated by the author).

Enclosures	Values of U thermal transmittance.		
	U_{exist} (W/m²·K)	U_{prop} (W/m²·K)	Reduction (%)
Façade	0.72	0.53	26.30
Roof	0.24	0.24	0.00
Joineries	5.70	2.84	46.70

envelope. The reduction of the enclosure's transmittance value is shown in Table 2 and is expressed in percentages [14].

The proposed intervention also considers replacing the stone finishing material with a double ceramic sheet of great durability and strength together with an internal mesh reinforcement [5]. The inner sheet of the enclosure is solved with a light sheet using rectangular stainless-steel profiles, laminated with gypsum board cladding and a paint finishing. Furthermore, to carry out the passive conditioning strategy, gaps are opened in the enclosure at the height of the lower false ceilings. Automated opening and closing doors are inserted. In the south façade, jet fans will be installed to allow adequate air circulation through the chamber and correct air distribution. This false ceiling chamber houses the aluminium panel plates containing the PCM, Rubitherm SP 21EK Hydrate Salt (Fig. 9). The properties of this material allow it to act as a radiant device by heating the air in the false ceiling in winter and cooling it in summer. In this way, it is possible to modulate the operating comfort temperature in the rooms below. Installing adjustable and automated vertical slats protect the window openings from the sun's rays along their route during the day. As for the interior, a false ceiling with a ceramic finish is proposed and, to avoid leaks in the chamber system, we suggest sealing using EPDM compression strips and EPS bands. This is complemented by a technical floor that incorporates the building's necessary installations.

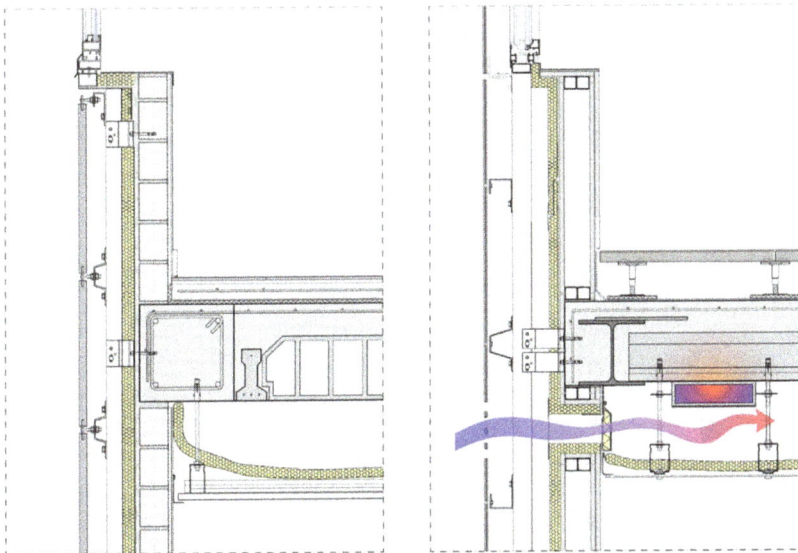

Figure 9: Operation process chart of the passive air conditioning system.

Table 3: Results of the energy demands of booth drawn up using HULC.

HULC Model	Energy demands			Reduction compared to the CTE reference building
	Annual	Winter	Summer	
	(KWh/m²·year)			(%)
Existing building	46.66	21.40	25.26	3.05
Proposed building	42.32	17.42	24.90	12.07
CTE reference building	48.13	16.34	31.79	-

6 CONCLUSIONS

Studies show that the use of ventilated façades on the east coast of Spain can reduce the overall environmental impact by 7.7%, based on a building's useful life of 50 years and the system's environmental amortization period of 30 years.

PCMs have shown to be fruitful in the field of architecture and are currently gaining ground. PCMs are used as a thermal storage system (TSS): they accumulate energy for use at different times, allowing to balance energy supply and demand, achieving an efficient use of the produced energy while taking advantage of thermal surpluses. Savings at peak times is another advantage of incorporating PCMs in buildings. Incorporating these materials to complement cold or heat production equipment allows compensating moments of high demand, increasing efficiency. The construction solution proposed in this paper is based on using PCM panels mainly as a thermal storage system, laid out in false ceilings and functioning as heat exchangers. The chamber's ventilation air flow can be controlled according to need.

Based on the studied ratios of energy gains per day of refrigeration supply compared to that of similar prior studies, the consumption of cooling energy for each building floor can be estimated to be 0.087 MJ/day·m², considering that 10 panels of 0.12 dm³ were installed per m². That is, a saving of approximately 0.90 KWh/ m²·year is achieved. Therefore, a 12.0% saving of combined heating and cooling demand is made.

The cost of the PCM materials, the openings produced in the building and its refurbishment means that the system's amortization period extends over time. Recent studies on the functioning of PCMs have shown that investing in this type of intervention is risky and the cost is high, sometimes so much so that it is not profitable or of interest to property developers.

REFERENCES

[1] Directive 2010/31/EU of the European Parliament and of The Council of 19 may 2010 on the energy performance of buildings.
[2] Ministerio de Fomento. (2013, 10 de septiembre). Orden FOM/1635/2013, de 10 de septiembre, por la que se actualiza el Documento Básico DB-HE «Ahorro de Energía», del Código Técnico de la Edificación. *Boletín Oficial del Estado*, n°219, pp. 67137–67209.
[3] Ministerio de Vivienda. (2006, 28 de marzo). Real Decreto 314/2006 de 17 de marzo. Código Técnico de la Edificación. *Boletín Oficial del Estado*, n°74, pp. 11816–11831.
[4] The Government of Spain. *Royal Decree 314/2006. Approving the Spanish Technical Building Code CTE-DB-HE-1*; The Government of Spain: Madrid, Spain, 2013.
[5] Fernández, A.E., Iribarren, V.E. & Iribarren, F.E., Energy efficiency of ventilated façades: Residential buildings, Alicante, Spain. *WIT Transactions on the Built Envi-*

ronment, vol. 171, WIT Press: Southampton and Boston, pp. 41–52, 2017. https://doi.org/10.2495/STR170041

[6] Abhat, A., Low temperature latent heat thermal energy storage: heat storage materials. *Solar Energy*, 30, pp. 313–332, 1983. https://doi.org/10.1016/0038-092X(83)90186-X

[7] de Gracia, A., Navarro, L., Castell, A., Ruiz-Pardo, A., Álvarez, S. & Cabeza, L.F., Experimental study of a ventilated facade with PCM during winter period. *Energy and Buildings*, **58**, pp. 324–332, 2013. https://doi.org/10.1016/j.enbuild.2012.10.026

[8] de Gracia, A., Navarro, L., Castell, A., Ruiz-Pardo, A., Álvarez, S. & Cabeza, L.F., Solar absorption in a ventilated facade with PCM. Experimental results. *Energy Procedia*, **30**, pp. 986–994, 2012. https://doi.org/10.1016/j.egypro.2012.11.111

[9] Echarri, V., Espinosa, A. & Rizo, C., Thermal transmission through Existing building enclosures: Destructive monitoring in Intermediate Layers versus Non-Destructive Monitoring with sensor son surfaces, *Sensors*, **17**, 2848, 2017. https://doi.org/10.3390/s17122848

[10] Pomponi, F., Piroozfar, P.A.E., Southall, R., Ashton, P. & Farr, E.R.P., Energy performance of Double-Skin Façades in temperate climates: A systematic review and meta-analysis. *Renewable and Sustainable Energy Reviews*, **54**, pp. 1525–1536, 2016. https://doi.org/10.1016/j.rser.2015.10.075

[11] Hasan, A. & Sayigh, A.A., Some fatty acids as phase-change thermal energy storage materials. *Renewable Energy*, **4(1)**, pp. 69–76, 1994. https://doi.org/10.1016/0960-1481(94)90066-3

[12] Kuznik, F., Virgone, J. & Noel, J., Optimization of a phase change material wallboard, *Applied Thermal Engineering*, **28** (11–12), pp. 1291–1298, 2008. https://doi.org/10.1016/j.applthermaleng.2007.10.012

[13] Suárez, R., & Fragoso, J., Estrategias pasivas de optimización energética de la vivienda social en clima mediterráneo. *Informes de la Construcción,* **68** (541), pp. 1–12, 2016. http://dx.doi.org/10.3989/ic.15.050

[14] Bienvenido-Huertas, D., Bermúdez, J., Moyano, J. & Marín, D., Comparison of quantitative IRT to estimate U-value using different approximations of ECHTC in multi-leaf walls, *Energy & Buildings*, **184**, pp. 99–113, 2019. https://doi.org/10.1016/j.enbuild.2018.11.028

[15] Código Técnico de la Edificación (CTE), Reglamento de Instalaciones Térmicas en los Edificios (RITE), ITC, 02.2.1.

[16] Iribarren, V.E., Garrigós, A.G. & Fernández, A.E., Energy rehabilitation of ventilated façades using phenolic panelling at the university of Alicante museum: Thermal characterisation and energy demand. *WIT Transactions on the Built Environment*, vol. 171, WIT Press: Southampton and Boston, pp. 3–15, 2017. https://doi.org/10.2495/STR170011

[17] Diarce, G. et al. Ventilated active façades with PCM. *Applied Energy*, **109**, pp. 530–537, 2013. https://doi.org/10.1016/j.apenergy.2013.01.032

A CASE STUDY OF GEOTHERMAL RESOURCES USE FOR THE INNOVATIVE AQUACULTURE FROM PERSPECTIVE OF SYNTROPIC DEVELOPMENT CONCEPT

LESZEK ŚWIĄTEK

West-Pomeranian University of Technology in Szczecin, Poland

ABSTRACT

Geothermal energy is developing with high progress to provide clean energy production standards at a world-wide scale. These projects are characterized with high risk level associated with drilling methods, resource existence, uncertain heat water temperature and its chemistry. The risk mitigation scenarios are crucial to avoid investment failure. Presented paper is a case study of geothermal investment in Trzęsacz, located in the Baltic coastline in Poland, where predicted heat water (38°C) was planned to be used for leisure, swimming and balneological purposes. The final effect of test drilling was disruptive. Thermal water has temperature 27°C and is not enough to fulfill needs of planned water park facilities and hot springs recreational proposals. The concept had to be revised. The amount of wasted water and embodied energy were recognized as a high entropy problem. In the spirit of syntropic development model, an idea to consume unwanted geothermal water and to treat it as useful local resource for aquaculture purposes was taken into consideration. That way the Jurassic Salmon Farm realization in Janowo in 2015, the first in the world salmon fishery based on geothermal resources, became an inspiration for future fishery deliberation, the fastest growing food sector globally. The Farm was realized 5 km from operating geothermal well, supported with EU funds and research programme led by West – Pomeranian University of Technology in Szczecin. The greenfield investment powered by renewable energy, based on biosafety and industrial ecology rules is an example of the 21st century bioculture. This one moved to urban areas may comply with broad sense to the city aquaculture, aquaponics or urban agriculture, with improvement of the risk reduction strategy in geothermal energy investments. This is the potential to be used by local communities, which can favor synergy effect on the way to regenerative design and syntropic development model.

Keywords: geothermal energy development, syntropy, sustainable aquaculture, regenerative design, risk mitigation.

1 INTRODUCTION

In the era of the fourth industrial revolution, the dominant part of the human population is concentrated in urbanized regions. Increasing cities are often conducive to concentration of economic capital, enable development and transfer of knowledge while struggling with the consequences of loss of natural capital. Both large metropolises as well as small and medium towns are affected by environmental problems, degenerative diseases of urban areas, imposing a negative ecological footprint in the natural environment both at the regional and global level. In order to minimize anthropopressure on the environment, discussions on healing the city's metabolism are underway, looking for sustainable development alternatives in the form of positive development, low-entropy city or syntropic development models. Newman *et al.* [1] reports 'The road to resilient cities requires finding ways of relating urban metabolism to practical, daily urban planning. We suggest that it can be done by examining urban metabolism in each of the urban fabrics. How can the materials and resources they use be regenerated and foster a mutually beneficial relationships between urban areas and the planet?'

© 2020 WIT Press, www.witpress.com

DOI: 10.2495/EQ-V5-N1-60-69

1.1 Characteristic of the syntropic development model

Promoting Nature-Based Solutions Pelorosso *et al.* [2] noticed: 'Entropy is a measure of the disorder, or waste of the city, and as such can be considered an indicator of the diversified impacts of the urban development on the biosphere. The entropy release of a city today is excessive because it overcomes Earth`s natural capacity of regeneration and threatens to destabilise the urban (human) civilization itself, which is causing it.' The counter force to increasing entropy is syntropy described by Hungarian biochemist Szent-Györgyi, Nobel Prize winner who postulates the existence of a force that causes living things to reach 'higher and higher levels of organization, order, and dynamic harmony' [3]. This way all living things can be characterized with the tendency to decrease entropy – in contrast to the tendency for inanimate matter to increase entropy (total equilibrium or total diffusion, producing a maximum of entropy and a minimum of free energy) [4]. Buckminster Fuller's student Baldwin [5] underlines: 'Bucky said that biology balanced entropy. Humans were the most powerful (known) antientropic forces of all, because we accumulate and purvey knowledge, adding local order to Universe in the same way that a plant synthesises air, sunlight, and soil nutrients into botanical life. Because anti-entropy is a double negative, Bucky called it 'syntropy'. Our purpose and duty as humans is to be syntropic'. The syntropic development model convergent to the positive development [6] is employing the recirculation economy, regenerative design and clean renewable energy production.

1.2 The syntropic approach to geothermal energy investment

Geothermal energy as one of renewable resources is developing with high progress to provide clean energy production standards world-wide. These projects are characterized with high risk level associated with drilling methods, resource existence, uncertain heat water temperature and its chemistry. Entropy as well as syntropy also applies to energy or investment economy. The risk mitigation scenarios are crucial to avoid investment failure characterized with high entropy level. The risk strategies should be a part of any urban development plans, feasibility studies and architectural design processes, where decreasing entropy level mechanisms should be supplemented with syntropic development models too. Geothermal projects are focused on energy aspects mainly, where the water is a medium that works as a heat supplier. Water is the environment of life. It's the one that cleanses, revives and regenerates. Water taken from geothermal projects can play a role as a vital resource for productive green areas restoration, municipal forest plantations or cultivation of urban agriculture and aquaculture supported with renewable energy in frames of urban water management. Together with the migration of people from rural areas to growing cities, the ideas of transferring agricultural production to densely populated urban areas are emerging. Design and cooperation in the implementation of the Jurassic Salmon farm in Janowo (West Pomeranian Voivodship in Poland), the world's first closed salmon farm based on geothermal sources and autonomous water recirculation systems (RAS, Recirculation Aquaculture System) became an inspiration to consider the future of aquaculture and geothermal projects [7]. This type of controlled fish farming is recognized as the fastest growing food production sector on a global scale, based on innovative technologies supported by extensive research programs and the interest of international investment capital.

1.3 Aquaculture, energy and urban food production

Since 2000, the increase in aquaculture production has been recorded at 7% per annum on a global scale. Currently, 50% of world fish and seafood consumption comes from controlled farms [8]. At the same time, on the example of a noble fish species, the dynamics of salmon consumption is characterized by a rapid growth on a global scale. Its consumption in 2015 was 1.2 million tons, while in 2050 it is to rise to the level of 5 million tons [9]. Within the European Union, it is estimated that 48% of natural fish stocks are affected by the overfishing and loss of the ability to regenerate and restore vital herds [10]. For this reason, the EU's Common Fisheries Policy assumes the promotion of environmentally-friendly aquaculture models. Compared to traditional fish farms in open water areas (e.g. in the Norwegian fjords), land based farms with RAS radically support the saving of water resources, eliminate the process of pollution and eutrophication – especially as sensitive marine areas as the semi-enclosed Baltic Sea basin. Fish farms planned under one roof provide stability and control of the breeding environment, maintain a high level of biosafety, reducing the risk of diseases, the appearance of parasites or predators (e.g. cormorants) depleting the size of the breeding stock. RAS technologies guarantee a constant level of production (comparable in quantity and quality in a weekly cycle throughout the year), eliminate the risk of fish escapement or theft occurring commonly in open ponds. The future of RAS is in autonomous large scale farms, based on integrated processing, local fodder plants, development of aquaponics and hydroponic gardening [11]. This leads to the creation of eco-friendly agro-machines - production and logistics systems such as environmental hubs that exploit the synergy effect, principles of integrated and sustainable supply chain, circular waste utilization and cascade energy consumption in order to strengthen the environmental efficiency of planned investments. Hence the urban locations of integrated fish farms and urban agriculture combined with the processes of revitalization of post-industrial areas (brownfield and greyfield investments) become an interesting research issue, especially when urban farms are often associated with a bottom-up social movement promoting the production of fresh local food and a network distribution system. Ecological technologies used in Jurassic Salmon can be a good example and inspiration for the implementation of such projects in urban areas with diverse functional systems, heterogeneous social structure or dense network infrastructure.

2 JURASSIC SALMON FARM IN JANOWO – A CASE STUDY

Initiated in 2010 by local investors, the project was implemented on the basis of EU funds under the Operational Programme 'Sustainable development of fisheries and coastal fishing areas 2007–2013' action 3.5 Pilot projects and on the basis of the research program 'Using geothermal saline water for fish farming and breeding'. Run by the West Pomeranian University of Technology in Szczecin under the supervision of professor Jacek Sadowski, with the procedural support of the municipalities of Karnice and Rewal. The analyzed case study is a greenfield investment, powered by renewable energy. A farm operating on the basis of strict biosafety regulations, industrial ecology, resource efficiency and responsible production was designed. It is the only breeding in the world in which salmon develops and grows in a very clean and microbiologically safe geothermal water from the Lower Jurassic period, and the third plant that produces from roe to an adult fishes under one roof [9].

2.1 The genesis of the project - the issue of ecological effectiveness

The plans of the investor associated with Trzęsacz (a former fishing village with the characteristic ruins of a Gothic church on the cliff of the central Baltic coast in the Rewal commune) assumed the implementation of a year-round holiday center in the historic palace complex. The revitalization of the seventeenth-century palace for the hotel function to be preceded by the implementation of the congress center with a complex of recreational pools on the outskirts of the palace park. The attraction of the entertainment part was to be an aqua park equipped with internal and external thermal pools with a total water surface area of approximately 1500 sq. m. An analogous water park powered by geothermal sources (water temperature about 36°C) operates in the German border town Ahlbeck – 3 km from Świnoujście. In the vicinity of the palace complex, the investor took a geothermal water intake – Trzęsacz GT-1 with a hole depth of 1 224.5 m.p.p. and with a flow rate of 180 m³/h and a working temperature of 25.4°C [12]. The project was co-financed from The National Fund for Environmental Protection (NFEP) subsidies, however, the obtained thermal parameters of geothermal water proved to be lower than assumed, although the chlorine-sodium saltwater obtained from the aquifer of Lower Jurassic has a mineralization of 13.5 g/l and have positive properties for balneological needs. However, the relatively low temperature of geothermal water did not guarantee the economic viability of planned, warm thermal pools, which should be additionally heated with the use of other energy sources. The local authorities of Lidzbark Warmiński town landed themselves in a similar situation, where the temperature of water obtained from the geothermal well was 21°C, which did not stop the implementation of thermal baths and the heating of swimming pools in the aqua park. The investor in Trzęsacz revised the investment plans and made a decision on other use of geothermal water. Heat recovery for the needs of the heating installation in the palace building and pumping geothermal water for the investor's agricultural plots located in Janowo (Karnice commune), distant 5 km from the water intake, were initially planned. The already partly chilled geothermal water is used for technological processes in fish farming. The investment was based on the innovative Danish aquaculture technology for closed land farms with water recirculation systems (RAS). As a result, geothermal water is used in an effective way not only for the purpose of obtaining thermal energy but also for feeding in a controlled environment of Atlantic salmon farming ecosystems.

2.2 Industry 4.0 and Nature 4.0 ecosystems on the example of the Jurassic Salmon farm

The central facility on the premises of the Jurassic Salmon Breeding Center established in 2013 is the modern, compact production building of the fish farm with a usable area of approx. 9,000 sq. m. The one-storey hall includes a number of technical rooms, including geothermal and sweet water treatment and iron removal stations, filter stations, sewage treatment plants, denitrification systems and sludge treatment [13]. The main building is occupied by separate rooms with different breeding tanks sizes for fish farms (Fig. 1). The production cycle lasts 20–22 months, starting from the spawning phase of eggs (imported by air from Norway or Iceland) through the growth of fry, smolt grow up to the rearing phase of adults reaching a commercial weight of 5–6 kg.

The building operates as a kind of a large, integrated device – a form of research equipment monitoring a specific environment of Atlantic salmon farming. The artificial ecosystems, type Nature 4.0, [14] were implemented, which reflect typical life cycles of fish, with five closed water circuits in two environments: cold fresh water and warm geothermal water from the 150-year-old Jury period. The breeding takes place in a hall only lighted with artificial light of variable intensity, including the backlighting of the interior of breeding tanks with colored LED light. The nozzles placed in cylindrical tanks regulate the velocity and direction of the water stream, in which the salmons swim upstream (Fig. 2). The investment area was equipped with an extended Building Management System (BMS) controlling many technological processes and environment, monitoring the signals of sensors testing the temperature

Figure 1: Salmon breeding tanks (5m diameter) – post-smolt cycle phase on the Jurassic Salmon farm. Picture of the author.

Figure 2: Salmon breeding tanks with a diameter of 12 m and a depth of up to 6 m – grow up phase on the Jurassic Salmon farm. Picture of the author.

of water and production rooms, oxygen dosing, stability of the biological deposit, or emergency states of particular circuits [13]. For the needs of the research program, the quality parameters of the water and its physicochemical parameters were analyzed, such as: pH level, ammonia, nitrites, nitrates, CO2, BZT5, CHZTcr, total phosphate and phosphorus content, within the tolerance range for fish with the highest environmental requirements [15]. Many automation, surveillance or monitoring solutions have been adapted to mobile devices with emergency notification applications, staff on duty around the clock. The technological project and related infrastructure were implemented based on the BIM system and a virtual spatial model made available in a virtual cloud. This makes it easier to manage a building at a distance, speeds up servicing and maintenance, which is a characteristic feature of the Industry 4.0 investment (Industrie 4.0).

Industry 4.0 is a collective term meaning the integration of intelligent machines, systems and introducing changes in production processes aimed at increasing production efficiency and introducing the possibility of flexible product changes. Industry 4.0 is characterized by the convergence of the physical and virtual world (cyberspace) in the form of cyber-physical systems (CPS), concerns not only technology, but also new ways of working and the role of people in industry. Industrie 4.0 (Germany) is a platform connecting representatives from various areas, including industry, policy, business and R&D aimed at standardization and increasing the security of network systems, creating legal frameworks, promoting research and innovation [16].

3 THE MODEL OF INTEGRATION OF CAPITAL AND KNOWLEDGE – CREATING INNOVATIVE SOLUTIONS

The realized pilot building of the Atlantic salmon farm Jurassic Salmon in Janowo received co-financing from the EU funds at the level of 5,8 million euro, with a total investment cost of about 10,4 million euro [9]. The investor's own contribution covered the implementation of a vast infrastructure providing the necessary utilities to the construction site. Infrastructural activities took place in the area of two neighboring communes: Rewal and Karnice. Local government authorities supported the investment especially in the planning and design phase. The Rewal commune has supported and streamlined the procedures for the implementation of the geothermal well and the water intake. One of the objectives was to distribute thermal energy from renewable sources to nearby recreation centers with positive operational parameters of the intake. The procedures enabling the execution of line investments in the commune road belt, in the area of the geothermal supply and return water infrastructure as well as the transmission of fresh drinking water from private wells of the investor were efficiently performed. Employees of the Commune Office were involved in constructively solving problems related to environmental procedures (e.g. water law) and in mediating in neighborly disputes of ownership and formal-legal interpretations of regulations. The rural municipality of Karnice favored the investment by supporting work on the local land development plan for agro-park in the area of approximately 13 hectares in the region of Janowo. The commune agreed to the extension of the municipal water supply network towards the village of Janowo, in order to supply the farm with fresh water in the initial phase of the investment. In cooperation with the district authorities in Gryfice, permission was granted to reconstruct the local road in order to ensure appropriate technical parameters of access to the farm and ensure continuity of supplies and cyclical collection of mature fish. The completed research program confirmed the correct functioning of the built-in pilot plant for fish farming. The geothermal waters used in Jurassic Salmon intensified fish production by shortening the breeding cycle and increasing their resistance to diseases, which reduced mortality rates in the herd scale. During the investment, apart from specialists from Poland, the project involved experts from Denmark, Iceland, Norway, France, Chile and Indonesia. At present, 21 people found employment in the Jurassic Salmon Breeding Center in Janowo. Weekly fish production has stabilized at the level of 12–15 tons, with a production capacity of up to 20 tons per week. The company was the first in the country to obtain the international certificate of The Aquaculture Stewardship Council (ACS) for sustainable fish farming [9].

The activities of Jurassic Salmon and scientists from West Pomeranian University of Technology of Szczecin during the implementation and commissioning of a modern fish farm in Janowo and the results obtained contribute to the development of aquaculture both on a national and international scale. The idea of controlled breeding is an example of the bioculture of the 21st century, which transferred to urbanized areas can be part of widely understood city aquaculture, aquaponics or urban agriculture. This will result in a different perception of Nature by urban residents in the near future than in today and foster the creation of an innovative economy.

4 REMARKS ON THE IMPLEMENTATION

The implementation of a technologically complicated facility and a complex research project in the unchallenged time frame, resulting from EU funding dates and procedures, imposes rigorous discipline in making investment decisions. The short schedule of design work, preparation and construction of the farm together with the necessary infrastructure did not allow for prolonging administrative procedures, entangling in legal formal disputes with unfavorable neighbors or stoppages during the execution of construction works. In the domestic conditions, it seems that carrying out such a difficult investment in a short time is easier in the open rural space than in densely populated, conflictogenic urban areas. Local authorities of small municipalities strive for such projects (environmentally friendly, generating new jobs), create systems of incentives and procedural facilitation of organizational support during the investment implementation period. However, there is a doubt about the ecological effectiveness of such activities. Is there no effect of some capital dissipation and loss of potential synergy of investments implemented in the greenfield formula compared to urbanized areas (with existing extensive installation and road infrastructure) where brownfield or greyfield investments can be implemented? In rural areas it is more difficult to implement new access roads, it is more difficult to maintain them during the winter period (snow clearing classes), it becomes cumbersome to provide rational terrain with a large disperse of recipients. Due to the distance of the farm there will be difficult access of service teams and experts in comparison to urban locations (availability of airports, accommodation base, universities and research institutes). The phase of investment preparation and design, with extensive administrative and environmental procedures and at the same time with short and strict deadlines for obtaining EU funds requires quick and efficient decision-making by both the investor and the project team. Transferring the design process to the virtual cloud, current (on-line) reporting of the process of introducing functional, technological and installation changes as well as operating on the 3D model of the planned investment significantly accelerated the decision-making process and the implementation of construction works. The anticipated stage of the investment made it possible to conditionally collect a previously made salmon hatchery and start breeding with the rigors of biosafety, practically during the construction period. Such activity enabled the launching of the breeding cycle and the early start of scientific and research work, as part of the dissemination of Industry 4.0 formulas.

Figure 3: Project of the Research Centre – Aquaculture Lab located in the postindustrial city fabric of Szczecin on the Odra River bank. (Muszalska A., Świątek L. ZUT Szczecin, 2018).

The Jurassic Salmon farm has confirmed the wider use of geothermal water than just for energy needs. After the tests, the adapted and treated Jurassic water turned out to be an excellent environment for salmon farming. In comparison with surface water resources, used at traditional farms, geothermal water, due to the hermeticity of the deep underground deposit, is free from chemical pollution or dangerous to fish parasites. RAS systems used on the farm have become a kind of filter for mineralized geothermal waters when they are reintroduced into the environment. After recovering thermal energy and mixing saline thermal water with fresh water in order to create a proper living environment for adolescent salmon, after re-purification (through mechanical filters and biological deposit) water from the farm returns to the environment through a drainage channel system, feeding the sweet and salt lake of Liwia Łuża. The presented solution can be a model example of the cascade use of geothermal energy to reduce energy losses, but also the practical use of thermal water resources for the needs of fish farming. In this model of water use, the costly second borehole that injects the cooled geothermal water back to the ground is eliminated. The described implementation is also a proof that the lack of positive temperature parameters of geothermal water obtained after the actual drilling has not necessarily led to discontinuation of further investment activities. Investor's determination, with the support of the scientific community and local self-government authorities, set the path for conducting an innovative investment in the 21st century food production sector, showing the method of reducing the level of risk in geothermal drilling and the possibility of using waters with low potential of thermal energy as an element of syntropic development model. Association of contemporary aquaculture with geothermal water acquisition opens the perspective of implementation of this type of investment in urbanized areas, equipped with a dense network of infrastructure, access to research centers, logistics centers and processing plants based on cascade energy consumption and industrial ecology formulas (Fig. 3).

5 CONCLUSIONS

The case described above can be characterized by a sentence: how to turn an investment failure into a success. The Jurassic Salmon Farm with its innovative aquaculture complex systems appears to be operating well, and plans are to expand fish processing plant to increase high quality food production. The compact fish farm, closed under one roof, is an example of an effective solution to the problem of a low enthalpy geothermal source and direct use of thermal water for fish breeding. The presented measure reduces the entropy level of geothermal investment, at the same time increasing its level of syntropy. Such a solution eliminates the need for a costly and troublesome reinjection well as a result of the process of thermal fluid treatment for fish farming and its environmentally neutral discharge into natural water reservoirs. The use of geothermal resource maximization was achieved. A compact farm can be used as a ready-made, repeatable component in a cascading geothermal energy utilization model. This type of investment requires large capital expenditure. However, it should be pointed out that the initial capital of the Jurassic Salmon farm was subsidized by European Union funds at the level of 75%, due to the pilot and innovative project. The peripheral location and ecological footprint of the greenfield investment, for which new infrastructure with extensive use has been implemented, raise some doubts. Both the implementation of water wise design principles (e.g. water footprint analysis) of land-based closed fish farms as well as the possibility to adapt aquaculture with RAS systems to the built environment requires further research. Have migrations of intensive agricultural production into urbanized areas in the 21st century right to exist, or are economically justified, it is the issue unlikely to be quickly resolved in the near future. But anyway the effectiveness of closed fish farm investment should be greater in urban areas, where intensive infrastructure, various media networks and users higher density exist. Geothermal energy effectiveness, current technological and logistics solutions enable the development of urban agriculture, with particular emphasis on aquaculture and aquaponics. Undoubtedly, this is a vast area of potential multidisciplinary research, planning or economic analysis. The city farms may create the effect of technological symbiosis inscribed in the hybrid cascade model of using geothermal resources (mix of thermal energy and thermal fluids) in greyfield or brownfield eco-industrial parks to generate the synergy effect. This will have an impact on the implementation of urban agriculture and building a positive development model towards a new syntropic city. However, the dissemination of urban agriculture intensified with renewable energy sources will demonstrate the bioculture development of city dwellers, the scale of social ecological intelligence and the type of the relationship with Nature, including the sublime Nature 4.0. This is potential for use by the local community, city authorities, academia and business representatives that can promote the synergy effect on the way to regenerative planning and syntropic development model.

REFERENCES

[1] Book: Newman, P., Beatley, T., Boyer, H., *Resilient Cities, Second Edition: Overcoming Fossil Fuel Dependence*, Island Press; Second Edition, Washington, p. 157, 2017.

[2] Book: Pelorosso, R., Gobattoni, F., Leone, A., Reducing Urban Entropy Employing Nature-Based Solutions: The Case of Urban Storm Water Management in: Papa, R., Fistola, R., Gargiulo, C. (eds.) *Smart Planning: Sustainability and Mobility in the Age of Change*, Springer International Publishing AG, p. 37, 2018.

[3] Journal article: Vargiu, J., Editor of Synthesis 1 (Introduction to article by Szent-Györgyi). *Synthesis* 1, 1(1), p. 14, 1977.

[4] Book: Scaruffi, P., *Thinking about Thought: A Primer on the New Science of Mind*, Writers Club Press, New York, p. 280, 2003.

[5] Book: Baldwin, J., *Bucky Works: Buckminster Fuller's Ideas Today*, John Wiley & Sons, Hoboken, pp.226–227, 1996.

[6] Book: Birkeland, J., *Positive Development: From Vicious Circles to Virtuous Cycles through Built Environment Design*, Earthscan, London, 2008.

[7] Journal article: Świątek, L., Akwakultura Miejska – model integracji kapitału i wiedzy w przestrzeni komunalnej – Natura 4.0. The City Aquaculture – capital and knowledge integration model in municipal development – Nature 4.0., *PUA*

Przestrzeń Urbanistyka Architektura, Vol. 1, Wydawnictwo PK, Kraków, 2017. doi:10.4467/00000000PUA.17.019.7137

[8] Book: Bregnballe J. A., *Guide to Recirculation Aquaculture. An introduction to the new environmentally friendly and highly productive closed fish farming systems*, Food and Agriculture Organization of the United Nations (FAO), EUROFISH International Organisation; 2015.

[9] Online sources: Jurassic Salmon sp. z o.o. Online, http://jurassicsalmon.pl. Accessed on: 12 Jul. 2016.

[10] Book: Neudörfer F. Sustainable Fish Aquaculture. in: Schultz-Zehden, A., Matczak, M., (eds.) *Submariner Compendium. An Assessment of Innovative and Sustainable Uses of Baltic Marine Resources*, Maritime Institute in Gdańsk, pp. 204–230, 2012.

[11] Book: Vinci, BJ., *North American Perspective on Land Based Aquaculture: Past. Present & Future*, The Conservation Fund Freshwater Institute, Shepherdstown, 2015.

[12] Online sources: Kowalski M. Land based salmons from Poland. Jurassic Salmon, Szczecin 2015, Online, http://www.ccb.se. Accessed on: 23 Jun.2016.

[13] Technical specification: Świątek, L., *Description for the Building Project of the production hall in Janowo for the company Jurassic Salmon*, AKCENT Pracownia Projektowa, Szczecin, 2013.

[14] Conference: Świątek, L., From Industry 4.0 to Nature 4.0 – Sustainable Infrastructure Evolution by Design. In: Charytonowicz J., Falcão C. (eds) *Advances in Human Factors, Sustainable Urban Planning and Infrastructure. AHFE 2018. Advances in Intelligent Systems and Computing*, vol 788. Springer, Cham, 2019. doi:10.1007/978-3-319-94199-8_42

[15] Research report: Sadowski, J., *The use of saline geothermal water for fish hatching and ongrowing*, ZUT Szczecin, 2015.

[16] Online sources: Piątek, Z., *Czym jest przemysł 4.0.?* Online, http://przemysl-40.pl. Accessed on: 28 Mar.2017.

SOCIAL NETWORKS OF SPORT AND THEIR POTENTIAL IN SMART URBAN PLANNING PROCESSES

RAQUEL PÉREZ-DELHOYO[1], HIGINIO MORA[2], RUBÉN ABAD-ORTIZ[1] & RAFAEL MOLLÁ-SIRVENT[2]
[1]Department of Building Sciences and Urbanism, University of Alicante, Spain
[2]Department of Computer Technology and Computation, University of Alicante, Spain

ABSTRACT
Information and data have become a new working tool for many disciplines including urbanism. Its incorporation into the field of urban planning is currently a process with great development potential. Within this context, citizens are one of the most important sources of data, providing relevant information for better smart city planning, adapted to their preferences and needs. In this sense, social networks are very powerful tools that city planners have to know directly from users the use they make of public space. It is clear that this information cannot be left out of the process of smart planning and design of today's cities. Specifically, this work focuses on the study of sport social networks and aims to determine which sport social networks offer the greatest potential for improving urban planning processes. To this end, the main existing social networks in this field are studied and, as a conclusion, the advantages and disadvantages that make these sports networks an opportunity to move towards smarter, more participatory and inclusive urban planning are discussed.
Keywords: citizen participation, citizen-centric urban planning, inclusive city, smart city, smart urban planning, social networks of sport, technology-aided urban planning.

1 INTRODUCTION
Urban planning is currently clearly defined by the need for a reformulation of the way in which the practice of this discipline are developed, in order to improve the way in which cities evolve to respond to the needs and preferences of their inhabitants [1]. In this reformulation, the Information and Communication Technologies (ICT) are the main lines and guidelines for change. The emergence of this type of technologies for sharing information, with a clear expansion in the last 10 years, has led to a new way of understanding and living the city. The significant boom in the use of portable smart devices such as mobile phones by citizens and the large number of Internet-based tools available have been the main causes of this change. Greater connection and social participation in the life of cities is made possible by these technologies [2]. This new way of conceiving and planning the city, more participative and inclusive, which applies ICT, responds to the idea of smart urban planning which, in turn, is included within the concept of smart city. A smart city allows citizens to interact with it in a multidisciplinary way and adapts in real time to their needs efficiently, through the innovative integration of infrastructures with smart management systems [3].

A good example of the application of these technologies to promote the participation of citizens is social networking. In the field of urban planning, the analysis of data retrieved from social networks makes possible new ways of addressing urban issues and intervening in urban public space in a much more flexible and immediate way compared to the greater rigidity of other traditional tools of urban planning. The ease of adding geographic information to these social networks has led to a substantial increase in the amount of information of this type available free of charge. That is why one of the most interesting challenges for current urban planning is to be able to analyse this information, which is generated by citizens on a voluntary basis, in order to find answers to questions such as the places visited or the patterns of behaviour and thus adapt to their preferences and needs.

© 2020 WIT Press, www.witpress.com
DOI: 10.2495/TDI-V4-N1-62-74

Specifically, this work focuses on the study of sport social networks and aims to determine which ones offer the greatest potential for improving urban planning processes. To this end, the study of the main social networks existing in the field of sport and the discussion of the advantages and disadvantages that make these sports networks an opportunity for urban planning is presented in this work. The main objective of this work is, therefore, to get to know the main characteristics that make some sports social networks valuable sources of information to aid decision-making processes on issues related to urban planning. This knowledge will enable city planners to move towards more smart, participatory and inclusive urban planning.

The rest of the work is structured as follows: in Section 2, the methodology is briefly described; in Section 3, a series of eight sport social networks is studied, and some previous related work is collected; in Section 4, a comparative study of these social networks is carried out and the advantages and disadvantages that make them an opportunity for urban planning are discussed; and finally, in Section 5, some conclusions are drawn.

2 METHODOLOGY

The methodology used to achieve the objectives consisted of three phases: (1) A first phase of selection of the most relevant social networks existing in the field of sport. This selection has been made on the basis of popularity criteria. (2) A second phase of study of the selected networks. For the particular study of each of the networks, both the web services they offer and the applications for mobile devices associated with these web services have been analysed. (3) A third phase of comparative study in order to discuss the advantages and disadvantages that make these networks an opportunity for urban planning. The study has focused mainly on networks that offer geographic information and geolocated spatial data.

3 STUDY OF THE MAIN SOCIAL NETWORKS OF SPORT

Sporting social networks relate the data of the routes made by the athletes with the factors that allow them to compete with each other, such as the time spent or the difficulty of the routes. This is the key principle on which the different types of sports social networks and their associated applications are based. In this section, a series of these social networks of sport, both their websites and applications, are analysed in order to be able to identify which of them offer the greatest opportunities to move towards smarter, more participatory and inclusive urban planning. More specifically, this work focuses on the sport of running.

3.1 Social network "Atleto"

Atleto [4] is a sport social network that focuses on the creation of a user profile based on the skills and habits of sports practice, in order to be able to show the user a selection of specific activities always in a radius close to their location. In addition, it is also possible to connect with other athletes who are using the application and share and assess each other the exercises carried out. The character of this social network is very basic (Fig. 1). In Europe it is not very well known. However, its use is very important in the United States which is its country of origin. In this location, examples of very different use can be found, from fitness classes to activities such as yoga. The low international penetration of this social network is also due to its recent creation in 2015. A newly created social network is more difficult to disseminate when well-established alternatives already exist. Regarding what is of interest in this research, Atleto does not have geolocated data generated by users that can be retrieved, so that, by not providing information of this type, the social network does not offer a priori interest for urban analysis within the context of this research.

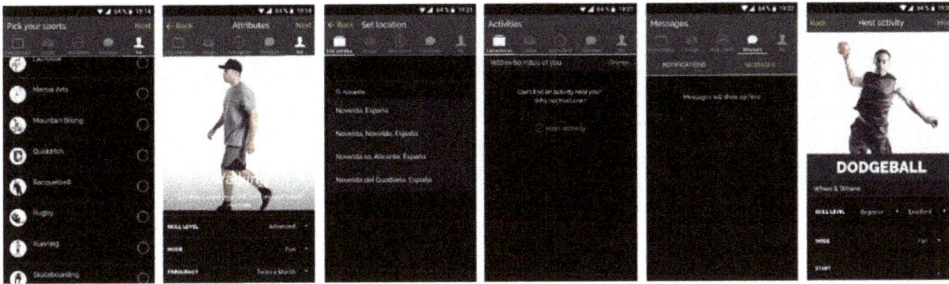

Figure 1: Social network "Atleto". View of some application screens *(Source: Atleto).*

3.2 Social network "Gotzam"

Gotzam [5] is a social network for athletes but also offers other services related to the organization of events. Its functionalities are related to organizational aspects of sporting events, such as event registration management, race timing, sports results management, as well as other services such as sports insurance, marketing or communication services.

This social network has no mobile application, so the web itself recommends the use of other applications, such as Runtastic, Runkeeper, Strava or Endomondo. However, it is possible to download the geolocated routes of the different events for analysis in GPX format. The potential of this social network, and the difference with other sport social networks studied, is the ability to inform users on the events that take place close to their location and that are related to the sporting activity they practice (Fig. 2).

3.3 Social network "Runkeeper"

Runkeeper [6] is a social network for sports monitoring that was created in 2008 and has belonged to ASICS since 2016. The principle of operation of this social network is an application that is based on the location of the sports activities of users in the cartography of Google Maps. However, the cartography is only used as a support to the training plans. In addition, a connection to other monitoring applications, such as Google Fit, is included, but without any social intention.

The data generated with this application are related to the physical characteristics of athletes (weight, sex, etc.). These characteristics are fundamental to improve their training and

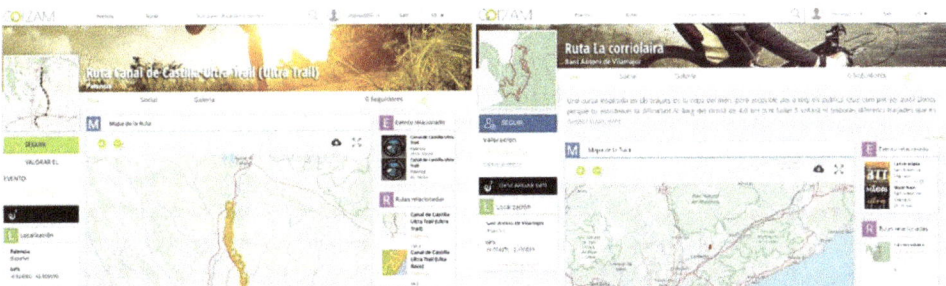

Figure 2: Social network "Gotzam". Examples of routes on the web *(Source: Gotzam).*

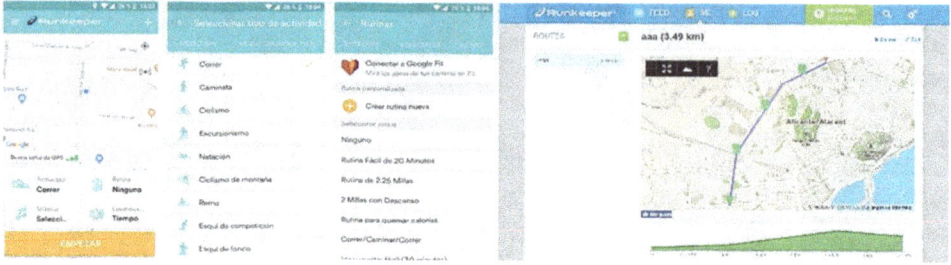

Figure 3: Social network "Runkeeper". View of some screens of the application and the interface to visualize routes on the web *(Source: Runkeeper).*

performance. In addition, users have the opportunity to make comments after their training and check whether they have achieved the proposed objectives.

Thus, the main attraction of this tool is the detailed analysis of statistics of the users' sporting activity. Although it is possible to compare the data obtained with that of other users, the potential of this social network is to create personalized training plans. In addition, the application has a gambling part, consisting of completing a series of challenges, which change over time in order to increase user motivation.

Regarding what is of interest in this research, it should be noted that it is not possible to obtain routes in geographical formats. It is only possible to visualize the routes and compare the times used (Fig. 3). Therefore, the social network does not offer a priori interest in the context of this research. With respect to the scientific literature, it has focused on analysing issues such as effects on health [7], and the impact of social sports networks in general [8,9]. The strong point of this application is to help the user to keep a precise control of his physical activity, so the studies carried out have to do with this issue.

3.4 Social network "Sports Tracker"

The social network Sport Tracker [10], created by Nokia in 2004, it was a pioneer in GPS tracking. It consists of a website and an associated application. The main functionality of this social network is to analyse the routes taken by athletes on a cartographic basis, offering data such as time, distance covered and other factors such as the speed reached and the difference in altitude of ascent or descent of the route.

Through the mobile application, the data are obtained and then analysed and presented on the web. Sports Tracker has a route search engine on map and it is possible to compare the records obtained with those of other users. Thus, it is possible to identify the points where sports performance can be improved (Fig. 4). In addition, once the sports activity has been carried out and a route has been registered, it is possible to share it in other social networks.

Figure 4: Social network "Sports Tracker". Search and route information on the web *(Source: Sports Tracker).*

It should be noted that geolocated data can be retrieved from this social network in GPX format. The greatest attraction of this application is its important development in recent years, which has led to the emergence of some specialized research [11].

3.5 Social network "Runtastic"

Runtastic [12] was created in 2009. Initially, this social network was based on the collection of data from athletes dedicated to sport in a playful way. Nowadays, it offers many more possibilities. It consists of a series of applications that share design and functionality and that form an ecosystem for measuring sports training.

The main features of Runtastic are as follows:

- Geolocated record of the sporting activities performed through the mobile application: from the geolocation platform of Google Maps, the user's activity is tracked and the result is shown when it is finished.
- Connection with the most popular social networks: through networks such as Facebook or Twitter, it is possible to establish comparisons with the exercise developed by other users, thus promoting a competitive aspect that helps greater use of the social network. This same functionality is also available on the own website.
- Motivation-based training plans: motivational messages and photographs of events are shown. Users are also given the opportunity to sign up for these events.

The social network Runtastic is very focused on the collection of data. As an important quality, its complete interface for the creation and management of routes should be highlighted. It is possible to filter routes based on a multitude of factors, such as distance covered, time spent and places visited. This facilitates a complete analysis of the routes (Fig. 5).

Runtastic also allows routes to be shared easily, in order to be able to make comparisons with the activities carried out by other users. There is a simple interface for searching for routes, both by geographical location and by route name, showing as a result very complete information.

In addition, it is possible to download the route map in vector format for import into a geographic information system. However, it must be taken into account that there is no API

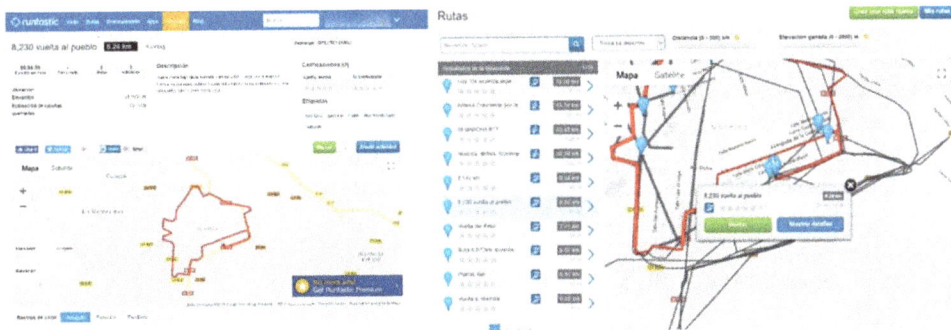

Figure 5: Social network Runtastic. View of different screens of the website: route visualization and search tool (*Source: Runtastic*).

(Application Programming Interface) to work with the data through the platform, and only manual actions can be carried out on each route obtained through the maps.

In regard to Runtastic, some scientific publications have been carried out [13].

3.6 Social network "Endomondo"

Endomondo is a social network that is part of a group of applications and tools belonging to the company Under Armour. The most important applications are MyFitnessPal; Under Armour (UA Record and UA HealthBox); and MapMy applications (MapMyRun, MapMy-Fitness, MapMyRide, MapMyWalk and MapMyHike), which record users' sports activities. These applications are interrelated in their features but are not integrated, i.e. there is no data interaction between them.

Endomondo allows users to create personalized training plans, follow maps and share training with other users [14]. The social network is linked to an application for mobile platforms that collects, at all times, the user's sporting activity when exercising. It focuses mainly on cycling and running activities but also includes other sports. Its user interface basically shows the routes covered by users, the times used and their progression. As soon as users complete the routes, they are presented with a results summary screen, in which the main parameters of length, distance covered, health effects and others parameters are described.

Athletes are also offered with the opportunity to share their impressions of the exercise performed, as well as to share the routes covered, through the most popular social networks. In addition to this, there are other options that are not free, with which it is possible to have access to other factors that affect the training, such as weather information, performance graphs or connection to external devices.

Complementing this core functionality, this application has a social interaction component, which is based on feedback through a notice board and the creation of events. It is possible to add friends in order to follow their trainings and compare the results. In addition, through events, in which users of the application from different parts of the world participate, it is possible to compete through the visualization of routes and basic training parameters.

The web version (Fig. 6) includes, in addition to the features listed above, the creation of personalized routes. It also includes the creation of personalized training plans in a more exhaustive way, although this option is not free of charge.

It should be noted that the most important variable that highlights in this social network is in regard to the user community and its integration with the most popular general social networks [15]. It is possible to share data on Facebook and Twitter, for example, so that any user can see all the information about the content of this application without having to download it. In this sense, it is worth highlighting the studies that have been conducted in this regard, which have shown that it is possible to fully analyse the data obtained from this social network [16].

As for the interest of this research, Endomondo allows users the free download of geolocated routes generated by users, so the social network offers a priori an interest for urban analysis in the context of this research.

3.7 Social network "MapMyRun"

MapMyRun [17] is a social network of sports monitoring belonging to the company Under Armour since 2015. It is based on the registration of sports activities from the location. Since

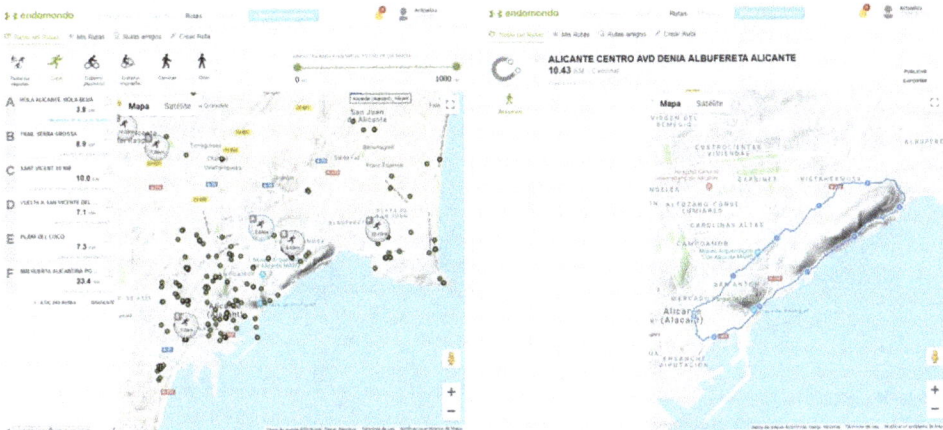

Figure 6: Social network Endomondo. View of different screens of the website: interface for searching and downloading routes *(Source: Endomondo).*

MapMyRun belongs to the same company as Endomondo, it has common characteristics with Endomondo, especially in the location aspects. It has a website and applications for the main mobile platforms.

The basic functions that it has are as follows:

- Training control: Through the use of Google Maps tools and the location system present in mobile devices, the application is able to obtain some data of exercise sessions that the user can use in order to analyse their performance.
- Recording and sharing of routes obtained with the application by GPS tracking: The creation of routes can be done in the application itself or on the website of the social network. Routes can be saved autonomously and shared immediately.
- Analysis of routes made by other users: The knowledge of these routes offers users the possibility of improving their own sports performance. Fig. 7 shows the information obtained when viewing a route made by another user.
- Creation of activity plans on the different sports disciplines: These plans can be complemented with other functionalities such as live training by a personalized trainer. However, these improvements are not always obtained free of charge, offering only limited functionality.
- This social network also has functions such as events and challenges, many of which are proposed by external service organizers. These functions also make it possible to compare sports activities with other users and share content.

Actually, there is little scientific literature on the social network MapMyRun. More research has been conducted on its sister social network Endomondo. However, an important finding is that, in most cases, the results obtained through the social network MapMyRun have been comparable to those obtained through the social network Strava.

Bearing in mind the similarities with Endomondo (both at company and operational level), it should be noted that MapMyRun focuses more on the person's own training than on the

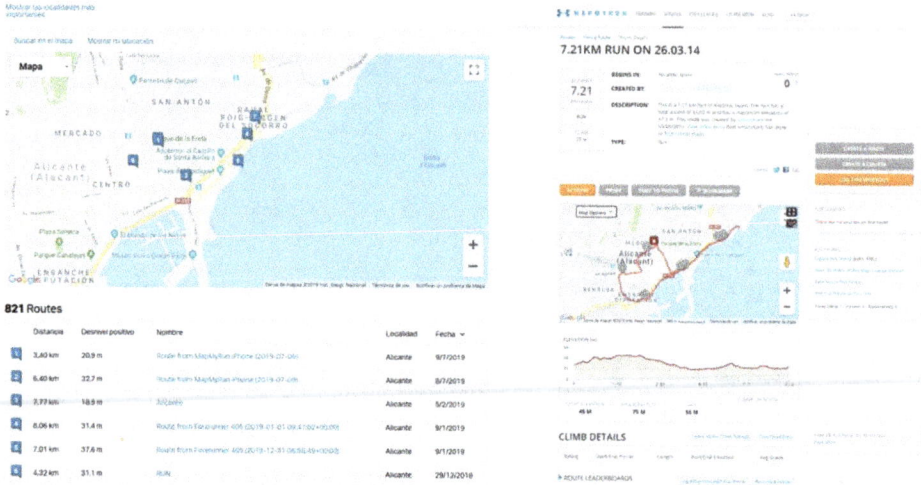

Figure 7: Social network MyMapRun. View of different screens of the website: interface for searching and downloading routes *(Source: MyMapRun).*

social side of the tool. With this it is possible to achieve better results in the trainings thanks to the existing functions.

In addition, it should be noted that the community of MapMyRun consists of a greater number of users who generate, therefore, a larger set of geolocated data downloadable for free.

3.8 Social network "Strava"

The social network Strava is currently one of the most popular and used in the field of sport [18]. In addition, there is an extensive literature related to this social network that consists of a website and its associated application [19,20].

The social network allows users, after selecting the sporting activity they are going to perform, to record the geographic data of the places they pass through while performing the activity and share them within the Strava community. In addition, it is possible to add multi-media components such as photographs related to the activities and it is also possible to share the activities through other social networks such as Facebook or Twitter. Another highlight is the possibility of creating a network of user friends for the purpose of comparison and monitoring of the activity performed.

From the website, it is possible to download and export the activities carried out by other users, regardless of whether or not they belong to the networks of friends. However, not all of these downloads can be made free of charge. So-called "segments" are available. Strava performs a series of calculations to identify the most popular "segments" where athletes practice their activities in the city. Also the "segments" can be created by the users themselves to motivate the competition, although with more exhaustive controls than in other social networks. These "segments" can be obtained from the existing ones in a previously selected geographical environment. In this way, they can be downloaded to the local computer, and analysed with geographic information systems (Fig. 8). In this respect, the variety of formats

Figure 8: Social network Strava. View of different screens of the website: interface for searching and downloading routes *(Source: Strava).*

in which information can be downloaded is highlighted, providing flexibility for analysis. It should be noted that Strava offers a very complete documentation of its API.

Also, the existence of the Strava Metro tool [21] should be noted, which analyses the data that Strava obtains through its application in order to provide information such as time spent on streets, and popular or avoidable routes. Obviously, with this tool at the disposal of professionals, it is possible to improve certain aspects of some disciplines, such as urban planning or that of transport engineering [22]. However, Strava Metro is not a free service.

On the other hand, Strava's social approach is clearly perceived. In fact, the interface of its application is divided into two different parts: the one corresponding to exercise and activity, in which it is possible to register all the sports activities carried out; and the one that offers the means to contact other users who carry out the same activities in nearby environments, with which it is possible to share different information about the activity being practised. In this way, Strava presents different characteristics with respect to other social networks studied. Not only does it allow the creation of routes that are registered on the Internet, but it also promotes community awareness with a clear desire for social interaction.

In the field of urban planning, quantitative and qualitative studies are as important, so taking into account only isolated data is not an option. This qualitative information makes it possible to gain a better understanding of citizens' needs, preferences and behaviour patterns.

4 COMPARATIVE ANALYSIS AND DISCUSSION OF THE ADVANTAGES AND DISADVANTAGES OF SPORT SOCIAL NETWORKS FOR URBAN PLANNING

Once these eight reference social networks have been described, this section discusses the advantages and disadvantages that make these sports networks an opportunity to move towards smarter, more participatory and inclusive urban planning.

With respect to the social network Atleto, in comparison with the rest of the networks studied, this network does not present a priori advantages to advance towards a smarter urban planning. The main reason is due to the lack of geo-location capacity of the sports activities that it includes, such as running, cycling and walking, among other activities. With the same criteria, the social network Gotzam does not offer users the possibility of registering routes through their own mobile application, nor tools to edit them. Gotzam only offers them the possibility of downloading other geolocated routes related to events. However, these geolocated routes do not contain user information, so it is not possible to know the actual routes that users take. Knowing the places visited by users of this type of social networks, through websites and/or their associated applications, is essential to improve the planning and design of cities. That is why it is determined that this social network does not present a priori advantages to move towards a smarter urban planning.

For the rest of the social networks of sport studied, and depending on the characteristics of geographical treatment of the data offered by them, the following considerations are highlighted.

It is evident that the support that each developer provides to social networks in the form of valid information that can be downloaded and analysed has a decisive influence. That is why certain social networks, such as RunKeeper, do not have advantages when it comes to studying the most popular places and routes chosen by citizens. The social network RunKeeper, both within the application and on the website itself, operates with a route system that is more adapted to the person doing the training than to the comparison between different trainings carried out by the users themselves. The routes obtained are isolated within the same training and it is not possible to obtain multiple routes from different users for analysis. This means that only all the routes of the user who performs them can be visualized. In addition, these routes cannot be downloaded, so their study is considerably difficult. This is the main reason why it is determined that the RunKeeper social network does not present any a priori advantages for urban planning either.

The social network Sports Tracker can be considered as a case similar to the social network Runkeeper, because it has historically been specialized in collecting data from users in order for them to check their performance. In addition, Sports Tracker has a major disadvantage, the rigidity when processing geolocated data. It is only possible to download data from complete routes and not from places where several routes can pass at the same time, so it is impossible to get a comprehensive view of the activity that occurs in a particular area of study.

With regard to the social network Runtastic and in relation to obtaining geolocated routes for analysis, it should be noted that, despite what has been said in previous sections, there is no API for downloading data. In this way, the Runtastic social network does not offer the user the possibility of downloading the data separately. It is obligatory to download the complete routes, without the possibility of classification of any kind.

The Endomondo social network allows users to download routes in GPX format by selecting each one of these routes from among those existing in a previously determined graphic window (specific geographical environment), choosing between the different sports activities such as running, cycling, walking, skating, etc. In addition to the geolocated routes that can be downloaded for free, this social network informs of the length of each route, as well as the number of times this route is followed. The routes are created by private users and they can have a greater or lesser number of followers. However, a large number of followers does not imply that the route has been covered by them. This data therefore represents a measure of the activity of users on the social network (on the Internet) but does not reflect the sporting activities that actually take place in the city.

Likewise, the social network MyMapRun allows users to locate existing routes in the environment of a given location. In addition to geolocated routes that can be downloaded in GPX format for free, this social network informs of the length of each route and the date it was created, allowing chronological studies. In addition, you can download a CSV file with information on the total ascent distance of the route and the elevation of each point of the route.

As in the social network Endomondo, the routes are created by private users and they can have a greater or lesser number of visits on the Internet. However, this data does not reflect the greater or lesser sporting activity that actually takes place in that specific environment of the city. This data only represents the activity of Internet users.

Both social networks, Endomondo and MapMyRun, since they allow users to download geolocated routes, are valid data sources to identify and analyse the routes and places preferred by citizens to practice their daily sports in the city. Therefore, both social networks have their interest in the field of urban planning, especially in decision-making processes, allowing the agents responsible for intervening in the city to take into account, as a priority, the needs and preferences of citizens. However, an important issue is the amount of data that these social networks offer free of charge. As an example, for the geographical area of the city of Valencia in Spain (longitude: −0.3773900; latitude: 39.4697500), MapmyRun offers more than 6,500 routes while Endomondo offers less than 800 routes. Consequently, the social network MyMapRun offers an important advantage in this respect.

In any case, both networks allow the analysis of particular routes created by users, although these are individualized systems that do not provide classifications, rankings or comparisons when the same routes are also travelled by other users. However, the social network Strava does establish these relationships. Once users finish their sporting activity, it is compared and classified with the rest of similar activities carried out by other users. Strava's large community, its classification tables and comparison "segments" give the social network a very advanced potential.

In the field of urban planning, it is very important to have data that reflect the greater or lesser sporting activity that takes place in a specific environment of the city. This knowledge of citizens' preferences facilitates the agents responsible for intervening in the city's decision-making processes. In this sense, Strava is a valuable source of data by allowing the downloading of "segments" in which a more or less elevated activity has been previously identified by passing several routes through that particular "segment". Thus, Strava offers free downloads of geolocated "segments" in GPX format and information on both the number of users who travel (although each user makes a different route) and the number of times that the "segment" is covered by the set of users. In addition, Strava offers other data on the difficulty of the "segment", its length, etc. The download of "segments", although not of complete routes, is permitted free of charge by selecting each of these "segments" from among those existing in a graphic window (specific geographical environment) determined previously.

5 CONCLUSIONS

In this paper, a series of eight relevant sport social networks in this specific field have been analysed. The study concludes that three of these social networks offer great advantages to move towards smarter, more participatory and inclusive urban planning. Of all the opportunities offered by these sport social networks, this paper has focused on those that offer services free of charge. The social network Endomondo and, more importantly, the social network MyMapRun, allow urban planners to have knowledge of the places and routes that urban athletes choose and prefer for their daily sporting activity. Based on this information, it is possible to characterize these places and establish criteria and priorities for intervention in the city, based directly on the citizens' own experience, i.e. their needs and preferences. On the other hand, the social network Strava provides additional information related to the intensity of the use of urban public space. With this data, urban planners can identify which of all the places preferred by citizens to practice sport in the city are really the most popular.

As a future work of this research group, it is proposed the application of geolocated data generated by citizens and retrieved from these social networks of sport for free, through the development of a series of case studies in real cities.

ACKNOWLEDGEMENTS

This work has been funded by the Conselleria de Educación, Investigación, Cultura y Deporte, of the Community of Valencia, Spain, within the program of support for research under project AICO/2017/134.

REFERENCES

[1] Landry, C., The changing face of urban planning: towards collaborative and creative cities. *Human Smart Cities: Rethinking the Interplay Between Design and Planning,* Springer, pp. 239–250, 2016.

[2] Hanzl, M., Potential of the Information Technology for the Public Participation in the Urban Planning. *Geoinformatics for the Natural Resources Management,* Nova Science Publishers: New York, pp. 475–498, 2009.

[3] Mueller, J., Lu, H., Chirkin, A., Klein, B. & Schmitt, G., Citizen Design Science: A strategy for crowd-creative urban design. *Cities,* **72**, pp. 181–188, 2018.

[4] Social network, "Atleto" website, www.atletosports.com/.

[5] Social network, "Gotzam" website, www.gotzam.com/.

[6] Social network, "Runkeeper" website, www.runkeeper.com/.

[7] Szark-Eckardt, M., Mobile applications as a tool conditioning health of young generation. *AIP Conference Proceedings,* **2040(1)**, pp. 070004, 2018.

[8] Martinez-Nicolas, A., Muntaner-Mas, A. & Ortega, F.B., Runkeeper: a complete app for monitoring outdoor sports. *British Journal of Sports Medicine,* **51(21)**, 1560–1561, 2017.

[9] Stragier, J., Vanden Abeele, M. & De Marez, L. Recreational athletes' running motivations as predictors of their use of online fitness community features. *Behaviour & Information Technology,* **37(8)**, 815–827, 2018.

[10] Social network, "Sports Tracker" website, www.sports-tracker.com/.

[11] Ferrari, L. & Mamei, M. Identifying and understanding urban sport areas using Nokia Sports Tracker. *Pervasive and Mobile Computing,* **9(5)**, pp. 616–628, 2013.

[12] Social network, "Runtastic" website, www.runtastic.com/.

[13] Antón, A.M. & Rodríguez, B.R., Runtastic PRO app: an excellent all-rounder for logging fitness. *British Journal of Sports Medicine,* **50(11)**, 705–706, 2016.

[14] Social network, "Endomondo" website, www.endomondo.com/.

[15] Vickey, T.A., Ginis, K.M., Dabrowski, M. & Breslin, J.G., Twitter classification model: The ABC of two million fitness tweets. *Translational Behavioral Medicine,* **3(3)**, 304–311, 2013.

[16] Fioravanti, A., Cursi, S., Elahmar, S., Gargaro, S., Loffreda, G., Novembri, G. & Trento, A. Visualizing and Analising Urban Leisure Runs by Using Sports Tracking Data. *Proceedings of the 35th International Conference on Education and Research in Computer Aided Architectural Design in Europe (eCAADe), vol. I,* pp. 533–540, 2017.

[17] Social network, "MyMapRun" website, www.mymaprun.com/.

[18] Social network, "Strava" website, www.strava.com/.

[19]   Hochmair, H.H., Bardin, E. & Ahmouda, A., Estimating bicycle trip volume for Miami-Dade county from Strava tracking data. *Journal of Transport Geography*, **75**, 58–69, 2019.

[20]   Sun, Y. & Mobasheri, A., Utilizing Crowdsourced data for studies of cycling and air pollution exposure: a case study using Strava Data. *International Journal of Environmental Research and Public Health*, **14**(**3**), 274, 2017.

[21]   "Strava Metro" website, https://metro.strava.com/.

[22]   Lee, K. & Sener, I.N., Understanding potential exposure of bicyclists on roadways to traffic-related air pollution: findings from El Paso, Texas, Using Strava Metro Data. *International Journal of Environmental Research and Public Health*, **16**(**3**), 371, 2019.

IS WALKABILITY EQUALLY DISTRIBUTED AMONG DOWNTOWNERS? EVALUATING THE PEDESTRIAN STREETSCAPES OF EIGHT EUROPEAN CAPITALS USING A MICRO-SCALE AUDIT APPROACH

ALEXANDROS BARTZOKAS-TSIOMPRAS, ELEFTHERIA MARIA TAMPOURAKI & YORGOS N. PHOTIS
Department of Geography & Regional Planning, National Technical University of Athens, Greece

ABSTRACT

In this paper, we evaluate different elements of the urban micro-scale environment in eight European capitals' downtown areas (i.e. Vienna, Copenhagen, Warsaw, Madrid, Brussels, Budapest, Athens and Sofia) to provide insight into inequalities in walkability benefits due to spatial distribution. To this end, we utilize MAPS-Mini, the brief version of Microscale Audit of Pedestrian Streetscapes to record and assess, at the street level, 15 walkability related items based on the Google Street View service. Our total sample consists of about 15.736 street segments/crossings, while for reliability analysis reasons, a second rater was employed to cross assess 10% of street segments per city. Results showed that Vienna and Athens had the highest (50.4%) and lowest (32.1%) overall walkability scores, respectively. Assessments were further combined with the population estimates of the European Urban Atlas 2012 dataset to perform equity analysis by estimating the distribution of average walkability scores among the population living downtown in the examined cities. In doing so, we used the Gini (G.) index and constructed Lorenz curve graphs. Our findings reveal a landscape of high inequality in downtown walkability distribution since all Gini coefficients were higher than 0.43. However, the inequality was greatest in Brussels (G. = 0.60) and lowest in Budapest (G. = 0.43). Additionally, we used spatial statistics tests (i.e. global and local Moran's I) to identify global and local patterns of walkability and population. The results indicated a highly clustered pattern of walkability across all downtowns and designated several clusters of uneven walkability geographies. Our approach sheds light on the application of active mobility strategies in different European cities, highlighting at the same time the need for further research to provide a clearer assessment of the spatial distribution of inequalities in social benefits and impact when designing sustainable urban neighbourhoods.
Keywords: active mobility, downtown, city centre, walkability, urban planning, equality.

1 BACKGROUND

Walkability enhances wealth, improves health, contributes to climate change mitigation, promotes transportation fairness and increases social capital [2]. To this end, equality in access to walkable built environments is an important element and a critical feature of a sustainable and resilient city [1]. Despite that, vibrant and highly walkable neighbourhoods often experience significant gentrification pressures [3] due to the high level of reinvestment and soaring housing prices [1]. Consequently, various vulnerable groups (i.e. immigrants and the poor) are displaced and excluded from the benefits of walkable urbanism and it is then when decision and policy makers need to analyze and evaluate the uniform spatial distribution of well-engineered pedestrian environments throughout the society.

Based on a micro-scale audit of pedestrian streetscapes, this study investigates the potential inequities in walkability spatial distribution among people living in downtown neighbourhoods throughout Europe. This research topic, although new within environmental justice literature, remains largely unexplored in the European context, as strong attention has been made so far towards unfair accessibility to amenities, such as food markets, recreation, transit stations, healthcare and education facilities [4,5].

DOI: 10.2495/TDI-V4-N1-75-92

There is various, although mixed, evidence that socially disadvantaged groups have unequal spatial access to high-quality and walkable urban environments and that the effects of the built environment on walking and physical activity are weaker within vulnerable populations. Adkins et al. [6] demonstrated that socioeconomically advantaged groups in the USA, living in activity-supportive built environments, walked two-fold times more and were more physically active than disadvantaged populations. To the contrary, a Belgian study [7] indicated that a neighbourhood's socioeconomic conditions did not interact with the association between walkability and physical activity. Bereitschaft [4] used the WalkScore® index and found significant inequalities in access to walkable communities among vulnerable groups in three US cities. Similarly, Riggs [8] demonstrated that walkable housing in the San Francisco Bay area is not inclusive, and specific minorities, such as blacks, are more likely to reside in car-dependent communities. In Buffalo, NY, Knight et al. [3] found that high WalkScore® values are concentrated in the highly gentrifying central and western parts of the city, while the poor tend to live in disproportionately low walkability and isolated districts [3]. Moreover, a Chinese and a Spanish study [9,10] reported that 15 min walkable neighbourhoods in Shenzhen are experiencing significant social inequalities, as spatial regression analysis showed positive correlations of adults and seniors with high walkability scores, but negative correlations of children and the nonresidential population. Respectively, in Madrid, the higher the socioeconomic neighbourhood status in the city, the lower the neighbourhood walkability index, while in gentrified and newly built areas this disadvantage was absent. Furthermore, a recent Scotland-based research [11] used a Geographic Information System (GIS)-based walkability index to demonstrate that higher area deprivation is not related to worse access to walkable areas.

Albeit all these walkability approaches are based on objectively measured macro-scale elements of urban form, such as population density, land-use mix, street network connectivity and transit/destination accessibility [12,[13], they do not consider the subtle differences of micro-scale urban environments. To this end, Bereitschaft [5] conducted a field survey in six streetscapes in Pittsburgh, PA, with equal macro-scale walkability (WalkScore®) values and revealed distinctive differences in the quality and maintenance of the urban environment between more and less disadvantaged communities. Neckerman et al. [14], based on a New York City study, reported that microenvironment disparities are shown only in walkable neighbourhoods and that the advantages of the macro-scale built environment can be eliminated due to differences in neighbourhood conditions. These findings underline the significant limitations of macro-scale walkability research, which fails to reveal the inequalities between the various sociodemographic groups living in the city [15].

Although micro-scale built environment attributes, such as sidewalk quality, crossings, lighting, aesthetics, etc., have been less studied than the macro-scale environment details in walkability research, they can be easily modified and might have a direct impact on population's physical activity levels [16,17]. For example, Cain et al. [18] found that micro-scale environmental features were correlated significantly and positively to objectively measured physical activity across all ages. However, data on micro-scale characteristics of the built environment are often lacking and are measured subjectively [17].

So far, several micro-scale walkability audit tools have been developed, including the following: (1) the Irvine-Minnesota Inventory [15,19]; (2) the SPOTLIGHT-Virtual Audit Tool [20]; (3) the Walkability Audit Tool of Centers for Disease Control and Prevention (CDC) [21]; (4) the Pedestrian Environmental Data Scan (PEDS) [22]; and (5) the Microscale Audit of Pedestrian Streetscapes (MAPS) [18]. The MAPS audit tool is one of the most widely studied instruments in active transportation and physical activity research [18] and has been

Table 1: Description of the pedestrian audit tool (Geremia and Cain [24]).

	Item	Theme	Answer/ points	Description
Segment	**S1***	Land use Type	0	Mainly residential or vacant space
			1	Non-residential (e.g. commercial, education, recreation etc.)
	S2*	Public parks or plazas	0	No access to park/plaza
			1	One access point to park/plaza along the route rated
			2	Two+ access points to park/plaza along the route rated
	S3	Public transit	0	No transit stop
			1	One transit stop
			2	Two or more transit stops
	S4	Public seating	0	None
			1	Yes
	S5	Street lighting	0	None
			1	Some
			2	Ample
	S6	Building maintenance	0	0–99% of buildings are well maintained
			1	100% of buildings are well maintained
	S7	Graffiti	0	Yes, the rated segment has at least one graffiti
			1	No, the rated segment is clean from graffiti
	S8	Cycling facilities	0	No
			1	Painted cycle lane
			2	Cycle lane separated from traffic with a physical barrier
	S9	Sidewalk	0	No
			1	Yes
	S10	Sidewalk maintenance	0	Poor maintenance or no sidewalk present
			1	Well-kept sidewalk
	S11*	Sidewalk buffer	0	No or no sidewalk present
			1	Yes or it is a pedestrian/car-free street
	S12	Overhead coverage (e.g. trees)	0	0–25% of the walkway length is covered or no sidewalk
			1	26–75% of the walkway length is covered
			2	76–100% of the walkway length is covered
Crossing	**C1_1**	Pedestrian walk signal	0	No
			1	Yes
	C1_2	Curb(s) ramp	0	No
			1	Yes, at one curb only
			2	Yes, at both pre- and post-crossing curbs
	C1_3	Marked crosswalk	0	No
			1	Yes

*Items are slightly differently defined in relation to the original MAPS-Mini guide.

systematically tested for use either in online or field observations [23]. MAPS is a 120-item tool developed in the United States for physical activity research. It has three alternative versions according to different research or community practice purposes – the MAPS Abbreviated [16], a 54-item version; the MAPS mini [25], a 15-item version ideal for practitioners; and the MAPS Global [26], a 123-item version suitable for international use.

Drawing upon previous MAPS research, we develop a short, online and massive segment-based walkability data collection method in eight different European downtown districts. Additional studies [27] from the United States have provided evidence that vibrant and walkable downtowns cause better mobility and health outcomes in the city. In this manner, we aim to quantify a micro-scale walkability indicator and, in turn, to reveal any potential walkability inequities among downtown residents. Although central urban cores are more walkable than suburban districts [1], we chose to focus on European downtowns as they often face significant socioeconomic segregation [28] and affordable housing issues [29], receive funds for costly regeneration and renewal projects [30] and play a critical role in daily city life. Therefore, this research seeks to answer the following questions:

1. Which downtown districts show the highest and lowest walkability performance?
2. Are walkable streetscapes of high quality distributed equally among downtowns?
3. Are walkability and population spatially clustered in downtown areas?

Findings from this study contribute mainly to the emerging European walkability literature as well as environmental justice research.

2 METHODOLOGY

2.1 Microscale Audit of Pedestrian Streetscapes (MAPS) – mini version

The brief (mini) version [25] of the 15-item Microscale Audit of Pedestrian Streetscapes (MAPS) tool [18] is simple, short and practical and facilitates our purposes for rapid, massive walkability data collection in many different city centres around Europe. Although MAPS-Mini was designed mainly for use by community groups and planning agencies, Sallis et al. [25] found that overall scores of the mini and the full MAPS versions are correlated positively at $r = 0.85$, and the relationship of MAPS-Mini scores with active transport is linear and positive for all ages. All the relevant manuals and guides for any MAPS version are available online at http://sallis.ucsd.edu/measure_maps.html (Accessed 10/8/2019). Table 1 describes each item of the audit tool of this study, as well as the contributed points of each item and answers to the overall score per street segment. The 15 items of the survey were selected in terms of their correlation with physical activity, attributed modifiability and consistency with guidelines for activity-supportive neighbourhood environments [25]. The most significant difference between our approach and the original MAPS-Mini guide comes up in the first, second, and eleventh segment items. In the first item, we gave one point not only to commercial street segments but to every route where more than 50% of the walkway's length was not residential or vacant space and included essential facilities for daily errands and leisure, such as schools, banks, food markets, cafes, bars, healthcare, etc. Next, in the second item, we decided to consider not only public parks but also plazas because they are vital spaces in almost every European city for children to play, and for many adults and the elderly to meet. Last but not least, in the eleventh item, we attributed 1 point to segments that were pedestrian streets.

2.2 MAPS-Mini data collection, scoring and inter-rater reliability analysis

Due to limited resources and the high geographic diversity of our case studies, we conducted online audits. Phillips et al. [31] found that the results of virtual audits, conducted by observers living in different geographic areas, show a high level of agreement with in-person MAPS audits. Each 'virtual' trained observer used Google's Street View service to audit the streetscapes; all were unfamiliar with the assessment areas (except the case of Athens, Greece). In each downtown district, we performed segment-level data collection, and we covered separately each side of the street, block by block, as well as each crossing on both sides of the street. The data management of the whole process was done in ArcGIS software version 10.3 (ESRI, Redlands, CA, USA), while the answers to the survey were recorded in polylines representing each side of the block, as well as their crossings. The polyline data were created by splitting the polygons of the European Urban Atlas 2012 dataset [32] that provides land use information at block level for numerous functional urban areas in Europe. Concerning the calculation of scores (0–100%) per each polyline record, we added up the total points of each rating, and then we divided the sum by 21. Also, for reasons of inter-city comparisons, we calculated the overall city indicators for every individual item of the audit tool (i.e. weighted by polyline length). These indicators helped us to create the total score of walkability attractiveness for each downtown area, as we multiplied each with the relevant points per answer and in turn divided their sum by 21. Finally, item-level reliabilities of a random sample of 10% of segments and their crossings per city were cross-assessed by a second rater. We used kappa statistics [33] for categorical variables and an intraclass correlation coefficient (ICC) [34] for overall walkability scores (continuous variables). According to Landis and Koch [33], the results of Cohen's kappa and ICC statistics indicate the following results about raters' agreement: 0.00–0.20: poor to slight; 0.21–0.40: fair; 0.41–0.60: moderate; 0.61–0.80: substantial; 0.81–1: almost perfect. This statistical analysis process was performed in SPSS version 23.0 (SPSS, Inc., Chicago, IL, USA).

2.3 Equity analysis in walkability and data sources

Equity pertains to the distribution of impacts, and this concept has been widely used in transportation planning decisions and transit research [35,36]. In this study, equity analysis in walkability refers to whether the distribution of walking-supportive neighbourhood designs is recognized as fair among the population residing in downtown districts. In practice, it is a horizontal equity concept where individuals should receive equal levels of walkability scores. To measure the level of equity in the distribution of walkability scores per each district, we used two classical measures. First, we employed the popular Gini coefficient [35,[36] which is a global statistical measure of dispersion widely used in economic analysis, as well as Lorenz curve graphs [36] to illustrate and explain the level of inequality. The Gini coefficient (G.) ranges from 0 to 1, where values near 1 indicate inequality and values close to 0 highlight equal distribution. The Lorenz curves (LCs) describe the cumulative distribution of benefits across the population. Thus, the diagonal line represents perfect equity, whereas the greater the area under that line, the higher the level of inequality in the distribution. To use these measures appropriately, we used two datasets. First, we utilized the population estimates of the European Urban Atlas 2012 (UA12) [32]. UA12 is a validated product created by an aerial interpolation GIS procedure; it is used by the European Commission's internal services to analyse population counts at high spatial detail [37]. However, in the case of Athens, Greece, we used the official population census 2011 [38], at block level as well. Second, for each polygon (i.e. city blocks) per downtown, we measured the average micro-scale walkability score. To do this, we converted all blocks to centroids, and from each of these points, we measured the mean walkability score of all overlapping segments at 250 m distance.

2.4 Spatial clustering analysis with local Moran's I

Spatial statistics measures can identify and map spatial clusters at global and local scales. First, we ran global statistical tests (i.e. global Moran's I) to determine in broad terms whether the patterns of walkability are clustered, random or dispersed. Second, since global tests do not report the locations or sizes of clusters, we drew upon local tests, such as univariate and bivariate local Moran's I spatial analysis [39]. Local Moran's index is a local indicator of spatial association (LISA) and has been widely used in geographical analyses to reveal hot and cold spots as well as to categorize them into spatial clusters or outliers [3,39]. Thus, to perform local spatial clustering analysis on micro-scale walkability scores, we used GeoDa 1.12.1.161. We calculated univariate local Moran's I statistics for each average walkability value per block (i.e. centroid) using a spatial weights matrix based on distance weights (250 m). Similarly, we applied bivariate local Moran's I statistics, which is a measure of spatial cross-correlation, to detect population clusters and walkability.

2.5 Study areas

The case studies were selected to include a geographically diverse group of European environments with different levels of economic development, heterogeneous urban mobility characteristics and various urban morphology features (i.e. population density). Also, the availability of recent (2011–2018) Street View imagery data was another primary criterion for the selection of cities (e.g. in German cities Google has an outdated Street View from 2008). Thus, the downtown areas of Madrid (ES), Athens (GR), Warsaw (PL), Budapest (HU), Vienna (AT), Brussels (BE), Copenhagen (DK) and Sofia (BG) were investigated. Table 2 depicts some essential characteristics of these cities, as well as some details about the downtown areas. All geospatial data regarding downtown boundaries were downloaded from OpenStreetMap.org, except in the case of Sofia, where the area denoted by Google Maps as the city centre was considered.

3 RESULTS

3.1 MAPS-Mini survey results

Overall, we rated and analysed 15.736 segments and 9.322 crossings (Table 2) from eight different European city centres. Table 3 shows the results for each item and the answers to the MAPS-Mini survey per examined city. In particular, Vienna showed the lowest share of mainly residential segments (9.78%), while Athens demonstrated the highest percentage (43.55%). Regarding the presence of segments with at least one access point to public parks

Table 2: Description of selected cities and their downtown areas.

City	Total population (2010) (millions)*	Population Weighted Density (inh./km²)*	Name of downtown district(s) (administrative)	Study area	Segments assessed MAPS-Mini		Crossings assessed
	Functional urban area			km²	Number	km	Number
Athens	3.92	14,73	1st District	7.2	4,505	307	2,678
Brussels	2.45	5,89	Pentagon	4.2	1,782	162	1,025
Budapest	2.88	5,13	District V & parts of Terézváros & Erzsébetváros districts	3.9	1,010	111	618
Copenhagen	1.67	4,34	Indre By	10.4	1,659	191	879
Madrid	6.25	18,8	Centro	5.2	2,512	222	1,418
Sofia	1.43	5,69	Център	7	2,288	238	1,612
Vienna	2.76	8,68	Innere Stadt	2.9	1,111	99	586
Warsaw	3.08	3,97	Śródmieście Północne and Południowe	5.9	869	125	506

*Data Source: European Commission, DG JRC, Urban Data Platform, https://urban.jrc.ec.europa.eu/#/en

or plazas, Copenhagen (23.42%), Vienna (20.22%) and Warsaw (20.33%) scored very high, whereas Sofia (7.7%) and Athens (8.97%) had the lowest values. Warsaw (23.62%) showed the highest number of segments with at least one public transit stop and Athens (8.39%) presented the lowest percentage. Copenhagen's (29.36%) and Budapest's (26.43%) downtowns displayed the highest proportions of streets with available public seating facilities (e.g. benches), but Athens and Sofia demonstrated markedly lower values than the other cities with 11.47% and 11.51%, respectively. Interestingly, Copenhagen (3.62%) highlighted the highest share of segments with no lighting, but this was the result of including in our analysis some segments of high length that belonged to a large park that did not have any lights installed. However, in Sofia, 3.17% of segments are unlit, while in Vienna the number of segments with ample lighting was the highest, at 29.78%. Furthermore, in Vienna (90.4%) and Copenhagen (83.13%), the buildings were found to be well maintained in most of their segments, while in Sofia (78.75%) and Athens (67.15%), the buildings were generally rundown. Graffiti and vandalism were a serious and pressing issue for three downtown areas, namely Sofia (77.13%), Madrid (70.02%) and Athens (68.19%), but in Vienna (9.09%) and Copenhagen (16.62%), this problem was far weaker. In the Athens city centre, we did not identify any segment with established cycling infrastructure, but Copenhagen showed the highest share of segments with protected bike lanes (27.68%) and Brussels (11.28%) demonstrated the highest percentage of segments with painted cycling lanes. Sofia's centre (5.12%) was the area with the greatest number of segments without any sidewalk present, whereas Madrid (0.55%)

and Vienna (0.73%) resulted in values under 1%. Additionally, Sofia and Athens had the worst level of sidewalk maintenance at 58.77% and 31.92% of their segments, respectively. Next, in Sofia (74.68%), we observed the highest number of sidewalk buffers, while to the contrary we identified the lowest share in Copenhagen (22.72%). Sofia (55.62%) and Athens (53.38%) also had the highest share of segments with minimum overhead coverage or shading, while Brussels (16.85%) presented the lowest value.

With respect to crossings, we identified the highest number of signalled crosswalks in Copenhagen (40%) and the lowest in Sofia (13.64%). Additionally, the highest number of crossings without ramps at any side of the curb as well as without any painted pedestrian crosswalk were found in downtown Athens at 59.83% and 82.31% of crossings, respectively.

Finally, in broad terms, Vienna's downtown demonstrated the best overall walkability score (50.45%), while Athens's and Sofia's central cities showed the lowest values with 32.08% and 32.39%, respectively. Copenhagen's and Warsaw's downtowns scored almost identical overall walkability scores with 48.93% and 48.48%, respectively. Madrid's central district ranked in the fourth position with 46.46%, while Brussels's and Budapest's urban cores resulted in the fifth and sixth ranks with 43.13% and 42.84%, respectively.

3.2 MAPS-Mini reliability analysis results

Concerning crossing reliability, almost all items had a near perfect agreement, with kappa statistics values ranging from 0.659 to 1. Specifically, the agreement between the observers in pedestrian signals item (C1_1) was perfect across all cities (Cohen's kappa > 0.971). Second, in segment reliability across all cases and items, we had moderate-to-perfect agreement. However, a fair agreement was indicated in Sofia's sample and at the cycling facilities item (Cohen's kappa = 0.395). Moderate agreement results (0.48< Cohen's kappa <0.60) were identified in four cities (Vienna, Warsaw, Budapest and Brussels) and three segment items (S1, S6 and S7). Finally, the ICC values for total scores were perfect in all areas (ICC>0.932). Results are described in Table 4.

3.3 Equity analysis results

Downtown Brussels showed the highest level of inequality (G. = 0.606) in walkability distribution. Similarly, the central cores of Warsaw (G. = 0.603) and Athens (G. = 0.593) demonstrated significantly high levels of walkability inequality. To illustrate this kind of inequality, according to LCs (Fig. 1), about 25% of downtown residents in these three cities receive almost 72% of the total walkability scores. In Copenhagen's (G. = 0.523) and Madrid's (G. = 0.496) central districts, the Gini indicators are lower, and about 25% of the population accumulates almost 63% of total walkability values. In Vienna's downtown (G. = 0.463), 25% of downtown residents take roughly 59% of total walkability scores. On the other hand, Budapest's and Sofia's city centres present the lowest Gini coefficients, with 0.441 and 0.435, respectively. The latter results, although they are the lowest Gini coefficients, mean that 25% of downtown residents receive about 56% of total walkability. To this end, Gini indicators and LCs from eight central districts around Europe underline relatively low horizontal equity in walkability distribution, since a significant part of high-quality walkable places is disproportionally available to a small number of residents.

Table 3: Indicators per MAPS-Mini item and answer (weighted by polyline length) and overall walkability scores per downtown district.

	Item	Answer	Athens	Sofia	Budapest	Brussels	Madrid	Warsaw	Copenhagen	Vienna
Segment	S1	0	43.55%	32.49%	19.34%	33.73%	20.06%	23.66%	42.98%	9.78%
		1	56.45%	67.51%	80.66%	66.27%	79.94%	76.34%	57.02%	90.22%
	S2	0	91.03%	92.3%	81.72%	85.82%	84.01%	79.67%	76.58%	79.78%
		1	2.62%	0.79%	2.2%	1.73%	2.32%	1.93%	3.67%	1.33%
		2	6.35%	6.92%	16.08%	12.45%	13.67%	18.41%	19.75%	18.89%
	S3	0	91.61%	91.58%	87.8%	90.5%	88.26%	76.38%	87.05%	87.24%
		1	7.38%	7.17%	8.42%	8.11%	9.93%	16.99%	11.28%	10.08%
		2	1.02%	1.25%	3.78%	1.39%	1.82%	6.63%	1.67%	2.68%
	S4	0	88.53%	88.49%	73.57%	85.29%	80.11%	77.03%	70.64%	77.79%
		1	11.47%	11.51%	26.43%	14.71%	19.89%	22.97%	29.36%	22.21%
	S5	0	1.09%	3.17%	0.67%	0.36%	0.00%	0.64%	3.62%	0.00%
		1	81.49%	86.67%	76.9%	78.64%	78.4%	76.12%	72.41%	70.22%
		2	17.42%	10.17%	22.43%	21.00%	21.6%	23.24%	23.97%	29.78%
	S6	0	67.15%	78.75%	44.66%	38.95%	28.82%	29.5%	16.87%	9.6%
		1	32.85%	21.25%	55.34%	61.05%	71.18%	70.5%	83.13%	90.4%
	S7	0	68.19%	77.13%	32.96%	36.7%	70.02%	46.28%	16.62%	9.09%
		1	31.81%	22.87%	67.04%	63.3%	29.98%	53.72%	83.38%	90.91%
	S8	0	100.00%	94.79%	92.2%	81.5%	98.66%	89.33%	67.02%	75.3%
		1	0.00%	1.69%	6.03%	11.28%	0.95%	3.54%	5.31%	9.89%
		2	0.00%	3.52%	1.76%	7.22%	0.39%	7.13%	27.68%	14.81%
	S9	0	2.57%	5.12%	2.19%	1.5%	0.55%	2.13%	2.16%	0.73%
		1	97.43%	94.88%	97.81%	98.5%	99.45%	97.87%	97.84%	99.27%
	S10	0	31.92%	58.77%	10.24%	6.28%	1.73%	7.09%	5.27%	1.81%
		1	68.08%	41.23%	89.76%	93.72%	98.27%	92.91%	94.73%	98.19%
	S11	0	56.74%	25.32%	72.18%	66.89%	36.24%	52.39%	77.28%	73.99%
		1	43.26%	74.68%	27.82%	33.11%	63.76%	47.61%	22.72%	26.01%
	S12	0	46.62%	44.38%	72.85%	83.15%	57.83%	67.04%	76.57%	69.46%
		1	27.51%	27.12%	11.9%	9.88%	22.14%	24.63%	14.95%	14.31%
		2	25.87%	28.5%	15.25%	6.97%	20.03%	8.33%	8.48%	16.23%
Crossing	C1_1	0	78.12%	86.36%	83.32%	79.55%	81.5%	65.93%	60.00%	75.11%
		1	21.88%	13.64%	16.68%	20.45%	18.5%	34.07%	40.00%	24.89%
	C1_2	0	59.83%	47.48%	5.3%	10.61%	3.04%	3.75%	2.7%	1.01%
		1	7.87%	16.75%	6.74%	8.15%	0.75%	1.8%	3.67%	2.16%
		2	32.3%	35.77%	87.96%	81.24%	96.21%	94.45%	93.63%	96.83%
	C1_3	0	82.31%	79.91%	68.6%	23.62%	33.61%	19.31%	42.26%	49.19%
		1	17.69%	20.09%	31.4%	76.38%	66.39%	80.69%	57.74%	50.81%
Overall score			32.08%	32.39%	42.84%	43.13%	46.16%	48.48%	48.93%	50.45%

Table 4: Segments, crossings and score reliability analysis results.

	Athens	Sofia	Copenhagen	Madrid	Vienna	Warsaw	Budapest	Brussels
Item			**Cohen's kappa statistics**					
C1_1	1	0.971	0.972	1	1	1	1	1
C1_2	0.972	0.923	0.806	–*	1	0.659	0.691	0.733
C1_3	0.99	0.98	0.983	0.988	0.946	0.823	0.924	0.948
S1	0.967	0.756	0.814	0.8	0.599	0.59	0.547	0.846
S2	1	0.936	0.92	0.953	0.678	0.646	0.643	0.942
S3	1	0.958	0.962	0.961	0.739	0.844	0.713	0.829
S4	0.938	0.939	0.917	0.937	0.683	0.888	0.789	0.896
S5	0.794	0.897	0.963	0.955	0.734	0.69	0.744	0.652
S6	0.973	0.921	0.802	0.883	0.598	0.714	0.571	0.485
S7	0.973	0.934	0.907	0.922	0.505	0.811	0.649	0.768
S8	–*	0.395	0.974	1	0.812	1	0.572	0.859
S9	1	1	–*	–*	1	1	–*	1
S10	0.896	0.942	–*	1	0.791	0.788	0.712	0.71
S11	0.986	0.945	0.944	0.964	0.675	0.834	0.691	0.954
S12	0.829	0.953	0.813	0.902	0.608	0.592	0.704	0.691
ICC	0.993	0.986	0.983	0.99	0.932	0.941	0.932	0.954

*Variables are constant

3.4 Spatial clustering results

Figure 2 displays the maps of average walkability values per block in eight central European cities. The Global Moran's I results suggest that all walkability scores are highly clustered in all city centres. The lowest value of Moran's (Fig. 3) indicator can be found in Warsaw ($I = 0.548$, $p < 0.00$), while the highest value is demonstrated in Budapest ($I = 0.906$, $p < 0.00$). The univariate local Moran's I maps (Fig. 3) describe the locations of statistically significant ($p < 0.05$) spatial clusters and outliers of walkability scores, and the results are classified into six groups: (1) high walkability clusters (scores are higher than expected by chance); (2) low walkability clusters (scores are lower than expected by chance); (3) high walkability outliers (scores are higher than expected by chance, but the scores of their neighbours are lower than expected by chance); (4) low walkability outliers (scores are lower than expected by chance, but scores of their neighbours are higher than expected by chance); (5) not significant (scores are equal to what is expected by chance alone); (6) Neighbourless (there is no neighbour, given the defined distance – 250 m – during the spatial weights matrix calculation).

In five cities, Madrid, Brussels, Budapest, Sofia and Athens, we have identified a large hotspot area of high walkability scores, surrounded by several smaller enclaves of low walkability. These high-quality pedestrian-oriented environments are located at the 'heart' of each downtown, and they have recently seen major urban transformations. For example, Brussels's and Madrid's [30] urban 'hearts' are experiencing large renewal projects, aiming to create extensive car-free and high-quality engineered central neighbourhoods. On the other hand,

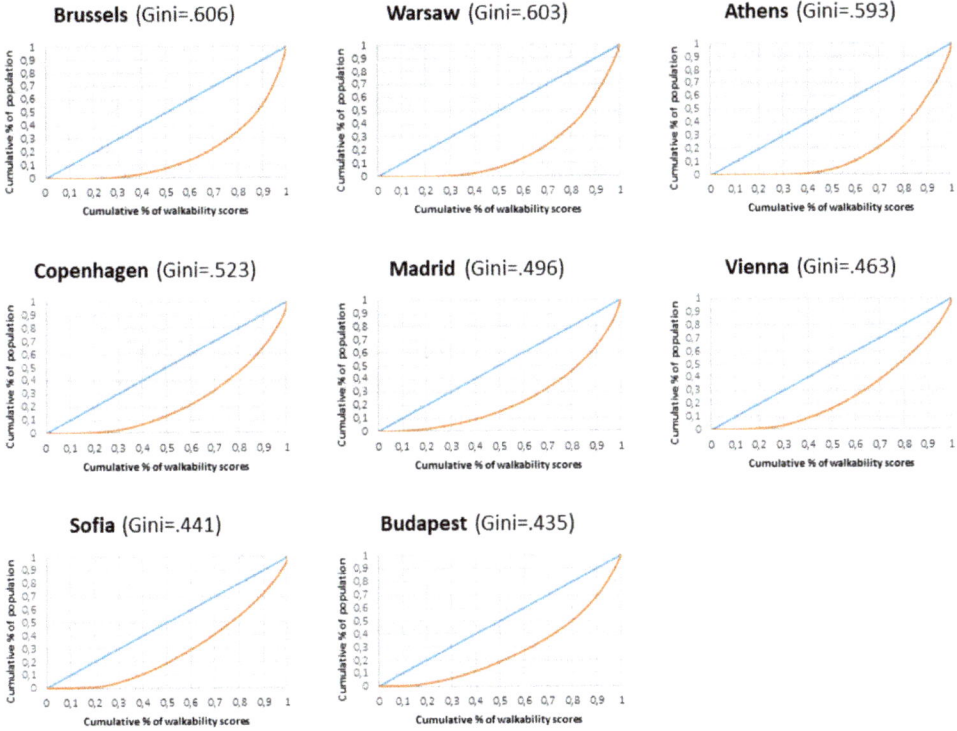

Figure 1: Gini coefficient values and Lorenz curve graphs per downtown district.

Figure 2: Spatial distribution of average micro-scale walkability scores per block.

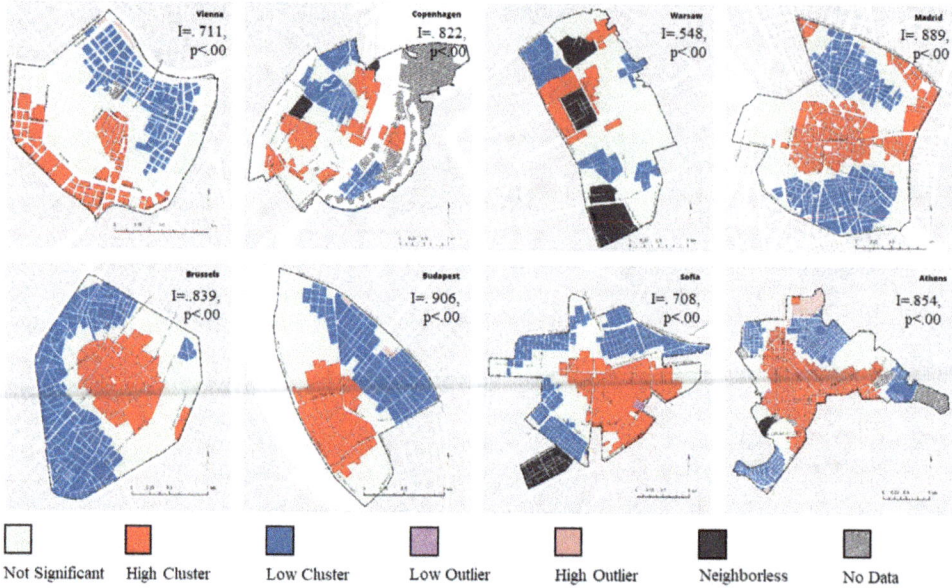

| Not Significant | High Cluster | Low Cluster | Low Outlier | High Outlier | Neighborless | No Data |

Figure 3: Spatial clusters/outliers of walkability per downtown (I = global Moran's index).

the cold spots of walkability are concentrated in districts that have chronically been excluded from urban policy shifts. For instance, in Athens neighbourhoods such as Exarcheia and Plateia Vathis, wherein various socially vulnerable groups reside (e.g. immigrants and the homeless), the streetscape walkability is lower than in more affluent areas (i.e. Kolonaki). In Vienna, the northern parts of the downtown seemingly show lower quality pedestrian environments in relation to southern and central parts that demonstrate extremely high-quality walkability features. In Copenhagen and Warsaw, we found several and dispersed clusters of low and high walkability enclaves, as well as many blocks that were classified as not significant. However, in Warsaw, due to huge block sizes, we had four neighbourless polygons. The number of blocks identified as low or high spatial outliers of walkability was negligible for all downtowns.

The bivariate local Moran's I maps (Fig. 4) display the spatial association of local populations living downtown and the micro-scale average walkability scores. In bivariate maps, the results are grouped into the following categories: (1) high population and high walkability; where a considerable proportion of downtown residents live in high-quality environments; (2) low population and high walkability, where in most cases these blocks pertain to parts of the city that encompass mainly large green, business or public areas of very low population densities; (3) high population and low walkability, wherein dwellers are more disadvantaged and live in lower quality places; (4) low population and low walkability, which often relate to military establishments or brownfield areas, with low levels of residences and non-activity-supportive built environments; (5) not significant or neighbourless, the explanation for these two groups is quite similar to those of univariate walkability cluster types.

In Athens (29.44%), Madrid (20.71%) and Budapest (20.48%), we identified the highest shares of areas with high walkability scores and low population (Table 5). These clusters are often located at the urban 'heart' (Fig. 4) and characterized as central business, tourism and/or government districts, wherein housing opportunities and practical affordability [29]

Table 5: Percentage of blocks area per bivariate (population and walkability) cluster type.

City	Bivariate cluster types (Local Moran's I)					
	Not sig-nificant	High population – high walkability	Low population – low walkability	Low population – high walkability	High population – low walkability	Neigh-bourless
Vienna	47.03%	8.34%	10.37%	15.94%	18.32%	0%
Copenhagen	44.49%	9.32%	9.98%	16.49%	15.95%	3.78%
Warsaw	50.35%	6.90%	3.70%	6.13%	13.15%	19.77%
Madrid	26.64%	6.94%	12.21%	20.71%	24.15%	9.35%
Brussels	34.95%	5.79%	13.12%	18.27%	27.87%	0%
Budapest	37.47%	6.69%	7.63%	20.48%	27.73%	0%
Sofia	29.44%	16.55%	16.62%	16.07%	14.54%	6.77%
Athens	28.46%	8.22%	17.19%	29.44%	15.05%	1.65%

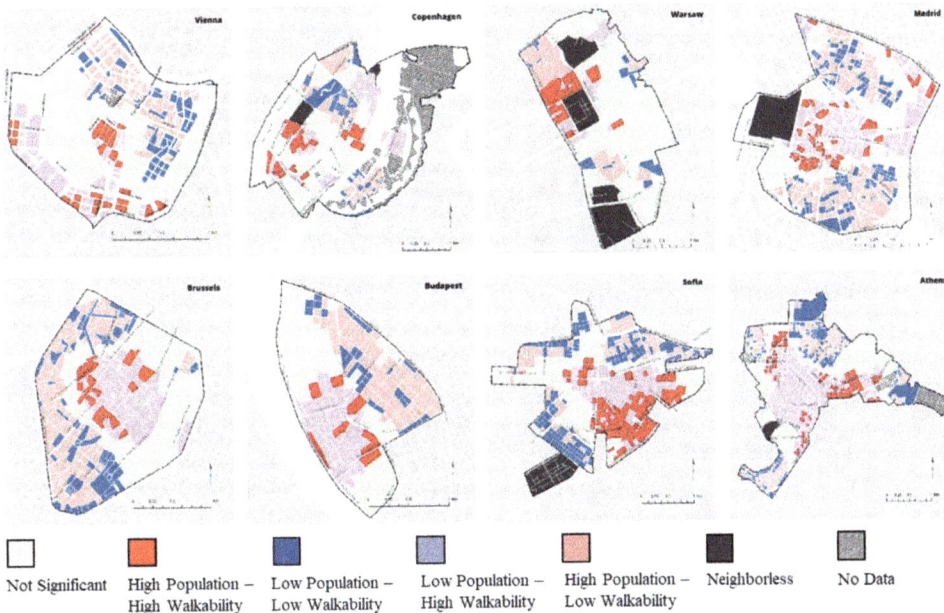

Not Significant — High Population – High Walkability — Low Population – Low Walkability — Low Population – High Walkability — High Population – Low Walkability — Neighborless — No Data

Figure 4: Bivariate local Moran's I cluster results (population and walkability).

mechanisms are slow to develop due to increasing rental prices driven by the walkability advantage [1]. This finding is consistent with other studies [3,8] that identified similarly alarming characteristics of high walkability in central places that exclude residents. Additionally, in Brussels (27.87%), Budapest (27.73%) and Madrid (24.15%), we highlighted more than 20% of areas with disadvantaged populations that reside in low-quality urban environments (Table 4). These locations are mainly residential and they are seemingly concentrated at the periphery of each downtown (i.e. the lower side of Christianshavn in Copenhagen, the Embajadores neighbourhood in Madrid, the Senne district in Brussels and the southwest Erzsébetváros area in Budapest). Thus, expansion of this analysis is required to understand if and to what extent this spatial disadvantage continuously affect people living at greater distances from the urban core. Interestingly, in Sofia (16.55%), we detect a high share of area with a high population and high walkability, which partially explains our previous result about Sofia's lower level in walkability inequality (Fig. 1). In all other cases, the high-high clusters were attributed to less than 9.32% of the total area. However, Athens (17.19%) and Sofia (16.62%) presented significant shares of blocks with low population and low walkability, which means that they could target these areas as potential locations for future urban renewal and reinvestments programs.

4 CONCLUSION

Everyone should have the right to live in high-quality and well-engineered sustainable urban environments with decent and accessible public transit options and multiple destinations nearby to walk safely and comfortably. To this end, several environmental and public health organizations have been calling for walkable urbanism to decrease the unsustainable impacts of car-based lifestyles and prioritize active mobility as a means to deal with the physical inactivity epidemic [2]. However, the spatial patterns of walkable neighbourhoods vary at different scales by socioeconomic context, and social inequalities are created [9]. In this study, we underlined the differences in measuring walkability between macro-scale and micro-scale approaches [5], and we applied a massive data collection process in eight European downtowns to calculate a micro-scale walkability index. We audited online 15.736 street segments and 9.322 pedestrian crossings, using the MAPS-Mini tool [25] and created average walkability scores per downtown block. The aim of this measurement was first to compare the differences in pedestrian microenvironments and second to find which downtown areas score the highest or lowest overall walkability attractiveness. Although the improvement of the examined microenvironment characteristics is a low-cost solution that requires short-term schemes [17] (i.e. graffiti removal), we identified striking differences between European central cities walkability attributes (Table 3). We concluded that Vienna and Copenhagen are the top walkable downtowns, while Athens's and Sofia's urban cores are the worst cases in terms of micro-scale walkability.

Next, we applied equity analysis in downtown micro-scale walkability values to answer the second question regarding the horizontal inequities of walkability distribution within European downtown areas. This question is crucial for urban sustainability; studies from the United States have demonstrated that downtown vibrancy is related to more population-level health and safety outcomes [27]. Therefore, we found that in all investigated European urban cores, walkability is highly unequally distributed among downtown residents, where in some cases (Brussels, Warsaw and Athens), one-fourth of the population receives about 75% of the total walkability scores. All Gini indicators were higher than 0.43, highlighting a landscape of great inequality that needs further socioeconomic and demographic investigation. This evidence is alarming for the urban planning scholarship since it proves that existing practice is not able to support inclusiveness in the intra-neighbourhood design, even at the most walkable part of a city, the downtown. However, similar Gini indicators in walkability are missing from other studies, and our results could not be compared to other regions.

Concerning the third question, we found that in all city centres walkability is highly clustered but at different magnitudes. We detected spatial clusters and outliers of walkability, as well as spatial clusters in terms of population and walkability altogether (Figs 3 and 4).

Cluster analysis helped us to demonstrate the highly unequal geographic distribution of walk-ability across the population in all downtowns, as we detected that the majority groups are concentrated mostly in deprived enclaves of low-quality environments at the periphery of the downtown. Nonetheless, high walkability clusters are often concentrated at the urban 'heart' of each downtown, but at the same time these high-quality environments can hardly be inhabited by many people. Notably, these 'uninhabited' and highly walkable clusters in seven of eight downtowns exceed 15% of the total area. Although we do not have available data about the housing markets and the sociodemographics for each region, we speculate that this result is influenced directly by increased rental prices and housing affordability/avail-ability issues [1] that in turn exacerbate and preserve inner-city social inequalities. Similar findings have been made by Knight et al. [3], where walkable and central blocks in Buffalo, NY, were highly clustered and gentrified. Furthermore, even in more developed cities with a high-quality of living, such as Vienna or Copenhagen, further efforts for renewal are needed, as we detected almost in all downtowns a significantly high number of isolated clusters where the population is high but the walkability poor. Nevertheless, since many renewal schemes are proposed or currently underway across European cities, their impacts on social cohesion should be fully acknowledged in the decision-making process.

Overall, this article has demonstrated how the concept of short US-based pedestrian audit tools can be operationalized to quantify and map micro-scale walkability attractiveness and equity in Europe. Our findings suggest that actions to improve walkability and the quality of downtown neighbourhoods should be entirely connected to socioeconomic and demographic justice targets. Evidence of highly unequal walkable downtowns across Europe reflects the broader socioeconomic inequities [28] of cities and critically hinders every path towards sustainable urban development.

However, certain limitations should be acknowledged. First, the boundaries of each down-town district are not commonly defined, and thus comparability between different areas is limited. Second, the online data collection process is based on Street View images captured at different time periods. Thus, observers provide walkability ratings that in some cases are outdated, but in other cases are quite recent. Third, the average walkability scores of blocks located at the boundary edges are biased, since ratings for segments and crossings outside the boundary are missing. Furthermore, the streetscape assessment process is human intensive, and it is difficult to implement at the metropolitan level. Last but not least, lack of fine-scale pan-European urban demographic data (i.e. on the elderly, children, youth, gender, etc.) and the socioeconomic conditions limits this research as well. Future researchers could expand and improve upon our approach in more downtown areas to cross assess and monitor the changing pedestrian microenvironment dynamics as well as to cross-correlate micro-scale walkability with travel behaviours and other sociospatial phenomena.

ACKNOWLEDGEMENTS

This research was co-financed by Greece and the European Union (European Social Fund) in the context of the project 'Strengthening Human Resources Research Potential via Doctorate Research' (MIS-5000432) implemented by the State Scholarships Foundation.

REFERENCES

[1] Gilderbloom, J.I., Riggs, W.W. & Meares, W.L., Does walkability matter? An examination of walkability's impact on housing values, foreclosures and crime. *Cities,* **42(PA),** pp. 13–24, 2015.

[2] Speck, J., *Walkable City Rules: 101 Steps to Making Better Places*, Island Press: Washington, DC, 2018.

[3] Knight, J., Weaver, R. & Jones, P., Walkable and resurgent for whom? The uneven geographies of walkability in Buffalo, NY. *Applied Geography,* **92**, pp. 1–11, 2018.

[4] Bereitschaft, B., Equity in neighbourhood walkability? A comparative analysis of three large U.S. cities. *Local Environment,* **22(7)**, pp. 859–879, 2017.

[5] Bereitschaft, B., Equity in microscale urban design and walkability: A photographic survey of six Pittsburgh streetscapes. *Sustainability (Switzerland),* **9(7)**, p. art. no. 1233, 2017.

[6] Adkins, A., Makarewicz, C., Scanze, M., Ingram, M. & Luhr, G., Contextualizing walkability: Do relationships between built environments and walking vary by socioeconomic context? *Journal of the American Planning Association,* **83(3)**, pp. 296–314, 2017.

[7] Van Dyck, D., et al., Neighborhood SES and walkability are related to physical activity behavior in Belgian adults. *Preventive Medicine,* **50(SUPPL.)**, pp. S74–S79, 2010.

[8] Riggs, W., Inclusively walkable: Exploring the equity of walkable housing in the San Francisco bay area. *Local Environment,* **21(5)**, pp. 527–554, 2014.

[9] Weng, M., et al., The 15-minute walkable neighborhoods: Measurement, social inequalities and implications for building healthy communities in urban China. *Journal of Transport and Health,* **13**, pp. 259–273, 2019.

[10] Gullón, P., et al., Intersection of neighborhood dynamics and socioeconomic status in small-area walkability: The Heart Healthy Hoods project. *International Journal of Health Geographics,* **16(1)**, p. art.no.21, 2017.

[11] Kenyon, A. & Pearce, J., The socio-spatial distribution of walkable environments in urban scotland: A case study from Glasgow and Edinburgh. *SSM - Population Health,* **9**, p. art.no.100461, 2019.

[12] Frank, L.D., et al., The development of a walkability index: Application to the neighborhood quality of life study. *British Journal of Sports Medicine,* **44(13)**, pp. 924–933, 2010.

[13] Bartzokas-Tsiompras, A. & Photis, Y.N., What matters when it comes to "walk and the city"? Defining a weighted GIS-based walkability index. *Transportation Research Procedia,* **24**, pp. 523–530, 2017.

[14] Neckerman, K.M., et al., Disparities in urban neighborhood conditions: Evidence from GIS measures and field observation in New York city. *Journal of Public Health Policy,* **30(SUPPL.1)**, pp. S264–S285, 2009.

[15] Koschinsky, J., Talen, E., Alfonzo, M. & Lee, S., How walkable is Walker's paradise? *Environment and Planning B: Urban Analytics and City Science,* **44(2)**, pp. 343–363, 2017.

[16] Cain, K.L., et al., Developing and validating an abbreviated version of the Microscale Audit for Pedestrian Streetscapes (MAPS-Abbreviated). *Journal of Transport and Health,* **5**, pp. 84–96, 2017.

[17] Sallis, J.F., et al., Income disparities in perceived neighborhood built and social environment attributes. *Health and Place,* **17(6)**, pp. 1274–1283, 2011.

[18] Cain, K.L., et al., Contribution of streetscape audits to explanation of physical activity in four age groups based on the Microscale Audit of Pedestrian Streetscapes (MAPS). *Social Science and Medicine,* **116**, pp. 82–92, 2014.

[19] Day, K., Boarnet, M., Alfonzo, M. & Forsyth, A., The Irvine-Minnesota inventory to measure built environments: Development. *American Journal of Preventive Medicine,* **30(2)**, pp. 144–152, 2006.

[20] Bethlehem, J.R., et al., The SPOTLIGHT virtual audit tool: A valid and reliable tool to assess obesogenic characteristics of the built environment. *International Journal of Health Geographics,* **13(1)**, p. art. no. 52, 2014.

[21] Dannenberg, A.L., Cramer, T.W. & Gibson, C.J., Assessing the walkability of the workplace: A new audit tool. *American Journal of Health Promotion,* **20(1)**, pp. 39–44, 2005.

[22] Clifton, K.J., Livi Smith, A.D. & Rodriguez, D., The development and testing of an audit for the pedestrian environment. *Landscape and Urban Planning,* **20(1-2)**, pp. 95–110, 2007.

[23] Zhu, W., et al., Reliability between online raters with varying familiarities of a region: Microscale Audit of Pedestrian Streetscapes (MAPS). *Landscape and Urban Planning,* **167**, pp. 240–248, 2017.

[24] Geremia, C., & Cain, K., MAPS-Mini, 2015. [Online]. Available: http://sallis.ucsd. edu/Documents/Measures_documents/MAPS-Mini%20Field%20Procedures%20%20 Picture%20Guide_090815.pdf. [Accessed 10 8 2019].

[25] Sallis, J.F., et al., Is your neighborhood designed to support physical activity? A brief streetscape audit tool. *Preventing Chronic Disease,* **12(9)**, p. art. no. 150098, 2015.

[26] Cain, K.L., et al., Development and reliability of a streetscape observation instrument for international use: MAPS-global. *International Journal of Behavioral Nutrition and Physical Activity,* **15(1)**, p. art. no. 19, 2018.

[27] Braun, L.M. & Malizia, E., Downtown vibrancy influences public health and safety outcomes in urban counties. *Journal of Transport and Health,* **2(4)**, pp. 540–548, 2015.

[28] Tammaru, T., Marcińczak, S., Van Ham, M. & Musterd, S., *Socio-economic segregation in European capital cities: East meets West,* Taylor and Francis Inc., 2015.

[29] Inchauste, G., Karver, J., Kim, Y. S. & Jelil, M. A., Living and leaving. housing, mobility, and welfare in the European Union, World Bank Report. [Online]. Available: http:// pubdocs.worldbank.org/en/507021541611553122/Living-Leaving-web.pdf. [Accessed 2019 8 5].

[30] Chaplain, C., The European cities leading the way in car-free living in a bid to tackle toxic air pollution, 2017. [Online]. Available: https://www.standard.co.uk/news/ transport/the-european-cities-leading-the-way-in-carfree-living-in-a-bid-to-tackle- toxic-air-pollution-a3658216.html. [Accessed 17 8 2019].

[31] Phillips, C.B., et al., Online versus in-person comparison of Microscale Audit of Pedestrian Streetscapes (MAPS) assessments: Reliability of alternate methods. *International Journal of Health Geographics,* **16(27)**, 2017.

[32] European Environment Agency, Urban Atlas 2012. [Online]. Available: https:// www.eea.europa.eu/data-and-maps/data/copernicus-land-monitoring-service-urban- atlas#tab-gis-data. [Accessed 10 8 2019].

[33] Landis, J.R. & Koch, G.G., The measurement of observer agreement for categorical data. *Biometrics,* **33(1)**, pp. 159–174, 1977.

[34] Shrout, P.E., Measurement reliability and agreement in psychiatry. *Statistical Methods in Medical Research,* **7(3)**, pp. 301–317, 1998.

[35] Bartzokas-Tsiompras, A. & Photis, Y. N., Measuring rapid transit accessibility and equity in migrant communities across 17 European cities. *International Journal of Transport Development and Integration,* **3(3)**, pp. 245–258, 2019.

[36] Delbosc, A. & Currie, G., Using Lorenz curves to assess public transport equity. *Journal of Transport Geography,* **19(6)**, pp. 1252–1259, 2011.

[37] Batista E Silva, F., Poelman, H., Martens, V. & Lavalle, C., Population estimation for the Urban Atlas Polygons, European Commision, JRC Technical Reports, 2013. [Online]. Available: http://dx.doi.org/10.2788/54791. [Accessed 10 8 2019].

[38] ELSTAT, Population Census 2011, 2011. [Online]. Available: http://www.statistics.gr/en/home/.

[39] Anselin, L., Local Indicators of Spatial Association—LISA. *Geographical Analysis,* **27(2)**, pp. 93–115, 1995.

Author index

WITPRESS ...for scientists by scientists

The Sustainable City XIII

Edited by: **S. MAMBRETTI**, *Polytechnic of Milan, Italy and* **J. L. MIRALLES I GARCIA**, *Politechnic University of Valencia, Spain*

Containing papers presented at the 13th International Conference on Urban Regeneration and Sustainability, this volume includes latest research providing solutions that lead towards sustainability. The series maintains its strong reputation and contributions have been made from a diverse range of delegates, resulting in a variety of topics and experiences.

Urban areas face a number of challenges related to reducing pollution, improving main transportation and infrastructure systems and these challenges can contribute to the development of social and economic imbalances and require the development of new solutions. The challenge is to manage human activities, pursuing welfare and prosperity in the urban environment, whilst considering the relationships between the parts and their connections with the living world. The dynamics of its networks (flows of energy matter, people, goods, information and other resources) are fundamental for an understanding of the evolving nature of today's cities.

Large cities represent a productive ground for architects, engineers, city planners, social and political scientists able to conceive new ideas and time them according to technological advances and human requirements. The multidisciplinary components of urban planning, the challenges presented by the increasing size of cities, the amount of resources required and the complexity of modern society are all addressed.

The published papers cover the following fields: Urban strategies; Planning, development and management; The community and the city; Infrastructure and society; Eco-town planning; Spatial conflicts in the city; Urban transportation and planning; Conservation and regeneration; Architectural issues; Sustainable energy and the city; Environmental management; Flood risk; Waste management; Urban air pollution; Health issues; Water resources; Landscape planning and design; Intelligent environment; Planning for risk and natural hazards; Waterfront development; Case studies.

ISBN: 978-1-78466-355-1   eISBN: 978-1-78466-356-8   **Published 2019 / 734**